PRACTICAL CASTING
A Studio Reference

Tim McCreight

BRYNMORGEN PRESS
Brunswick, Maine

ACKNOWLEDEMENTS

This book would not have been possible without the support and skill of many people. The manuscript was reviewed and improved by Fred Woell, Douglas Legenhausen, Linda Weiss-Edwards, Michael E. Moser, Dakin Morehouse and Steve Brown, along with hundreds of students who taught me so much.

My wife Jay has played a vital part in this book; her advice, good judgement and moral support are reflected on every page. Our children, Jobie and Jeff, are a boundless source of enthusiasm to us both.

Thank you, thank you.

Brynmorgen Press

Copyright 1986, revised 1994
Printed in color 2019

For ordering and contact information, visit
www.Brynmorgen.com

Libary of Congress Catalog
85-073045

ISBN: 978-0-9615984-5-7

Printed in Hong Kong

Contents

INTRODUCTION

When I wrote *The Complete Metalsmith* in 1982 I envisioned a series of books that would share the same format. This is the second step toward that idea. I hope it retains the popular aspects of the first book and introduces some improvements. One of the ideas behind this series is to divide information into easily assimilated pieces that are arranged in a straightforward practical sequence. The format is largely visual, and has a bias toward the simplest, cheapest way to achieve a specific result.

This volume, like its predecessor, is printed on heavy paper and uses a plasticized cover. The spiral binding will allow the book to lay flat on a workbench. Optional index tabs are provided at the back of the book so you can create your own chapter dividers if you would find them helpful.

The first five chapters deal with lost wax investment casting, the method most commonly used in jewelrymaking. The process is divided into its several aspects, and each is discussed fully. The remaining chapters describe alternate casting techniques; some are simpler than investment casting and some are more complex. A glossary is provided at the end of the book for quick reference. This book makes no attempt to cover related aspects of metalworking such as soldering and finishing. It is because of this singleness of purpose that *Practical Casting* can offer the complete coverage it does.

While most of the infomation in this book can apply to casting on any scale, bear in mind that it is written as a classroom text for situations where an instructor is present and volume is limited. It is not intended as a resource for industry, where other safety requirements may exist. For help in this area, contact your state office of Occupational Safety and Health Administration (OSHA) or the Industrial Safety Division of your state Labor Department. It is your responsibility to protect yourself and those around you.

OVERVIEW

Considered historically, casting is not a specific technique but a logical extension of the process of refining ore and consolidating nuggets. We can imagine that when the first metalsmiths found their material in the ashes of a fire, they discovered that it retained the shape of the ground on which it was spilled. All subsequent technology in casting has developed from this beginning.

When a worker pours an ingot as the starting point for fabrication, it is logical to cast a shape that is close to the desired end product. When making wire we cast a rod, and so on. Somewhere in ancient history, a metalsmith realized that the ingot could be poured to an almost finished shape. With this discovery the ingot mold became the object mold, and casting as an art form was born.

It appears that this revolutionary discovery was made independently by several cultures. We know of highly sophisticated bronze castings made in China about 7000 years ago. Not long after that, the ancient Egyptians evolved highly developed casting skills, creating stunning work that has inspired generations of metalsmiths. Around 500 BC, Greek smiths developed the blast furnace, in which a jet of forced air was directed at iron ore. This important development produced consistently better metal at a relatively low cost. In one sense, this marked the beginning of the Industrial Revolution.

Compared with most enterprises, casting has changed very little in the last 4000 years. The materials have been improved and the scale has grown, but the process is virtually intact. Also intact over the years is the terminology.

Mold (or, in England, *mould*) is the negative impression into which a softened or fluid material is pressed, poured or injected to achieve a pre-determined shape.

A mold is an intermediary step between two objects: the original pattern and the finished casting. The material and complexity of the mold depends on each of these elements. A complex pattern will dictate certain kinds of molds and rule out others. The melting point of the material being cast also restricts the choice of mold material.

Some molds can be separated for removal of the casting. These are called *piece molds*, They may be made of only two pieces, or may be very complicated and involve a dozen or more parts. Some piece molds use flexible materials to provide increased versatility. Others are made in one piece which must be broken apart after being filled to retrieve the cast piece. These are called waste molds, because the mold must be destroyed, (i.e., *wasted*) with each use.

The pattern or model is an exact image of the object being produced. The model material is chosen to suit the mold, the casting method and the intentions of the modelmaker.

In manufacturing, casting is one of many processes used to duplicate a shape. Casting takes its place among other processes such as blanking, drawing, stamping, spinning and so on. The manufacturer selects a production method that will create the highest yield of the best product at the minimum expense. Casting often fills these requirements, with new applications being found every day. Techniques originally developed for dental work were taken up in the 1940s by the jewelry trade, and today, in addition to jewelry, hundreds of thousands of small parts such as valves and machine components are being cast. As the manufacturing community turns its attention to casting, developments and refinements occur with great speed. More research is being done on casting each year now than was done in the whole decade of the 1950s.

Without meaning to diminish the importance and excitement of moldmaking and casting, it must be noted that the peak of creativity is in making the model. Most of the characteristics of the final piece—its shape, proportion, contours, and usually its surface—are resolved here. The success or failure of moldmaking and casting is defined as the degree of similarity between the original model and the finished casting. Every artisan is open to the value of "spontaneous design" (happy accidents) but control of technique is important to good design.

To borrow a concept from the noted craft author David Pye, the model is a "storehouse of the time and talent" of its creator. Through the technical expertise of the moldmaker and the caster, the skills of the designer and modelmaker can be realized. In fact with a reusable mold, they can be realized in great number. Technical skill, no matter how great, cannot improve deficiencies in the original model. The best efforts in casting a mediocre model will result in a mediocre duplicate.

Division of labor is almost as old as metalsmithing itself. Ancient Egyptian paintings show apprentices doing the busywork of their accomplished masters. In that and most other cultures it was generally true that a master achieved skills and status by working through the levels of the workshop. Though the master no longer performed the menial tasks of the shop, knowledge of them came from personal and usually long experience. In later cultures, and most especially since the 19th century industrial revolution, roles within a large casting shop have been hired out as needed. It is possible therefore to find a modelmaker who is only vaguely informed of the subsequent processes needed to convert that model into a finished work.

If there is a premise to this book it is that these subdivisions have little value, and are perhaps detrimental, to the artisan. For the jewelrymaker working on limited production or one-of-a-kind pieces, it is common to work a piece under one's own hands from start to finish. In a highly specialized world, this has an emotional appeal and a psychological reward.

The shape, scale and imagery of a model have more to do with design (an aesthetic topic) than with casting (a technical concern). A study of design is as fascinating as it is important but it is outside the scope of this book.

A sense of design—knowing of what shape to make—is both vital and elusive. Good design is a rare alchemy of perception, inspiration, diligence, and response to materials. For our purposes, we must be content to encourage diligence, reward perception, hope for inspiration and turn our attention to technique.

CHAPTER 1

MODELS

The technique most widely used to cast jewelry today is the lost wax investment mold process. The ability of this process to produce consistently clean castings with a minimum of equipment accounts for its wide popularity. The next five chapters describe this process in detail.

With a relatively small investment of time and money, any metalsmith will be able to master the process of lost wax casting. It should be noted, however, that not everyone has the space, temperament or need to set up a full-scale casting studio. An online search will quickly link companies that are in the business of casting the models of independent jewelers. For metalsmiths whose limited volume does not justify a large casting, setup these companies provide a valuable service.

The Lost Wax Process

1 A model is made of wax or another completely combustible material.

2 The model is mounted on a wax rod, called a sprue.

3 The sprued model is mounted onto a base and fitted with a watertight open-ended cylinder called a flask.

4 A plaster-like material called investment is mixed to a creamy consistency and poured over the model, filling the flask. Steps are taken to remove bubbles from the mix.

5 The investment is dried and then burned out (heated in a kiln) to remove all traces of the model. Lost wax, get it?

6 While the mold is still warm from burnout, molten metal is poured or forced into the mold, where it assumes the shape of the original model.

7 After brief cooling, the mold is quenched in water, which breaks it open and releases the casting.

Making a Model

The first step in making an investment casting is to create a model or pattern, that is an exact image of the object to be cast. The aesthetics of a piece—its shape, thickness, surface, and texture—are intimately bound up with the material being used. Experimentation and experience are the best teachers of these important factors.

Wax is the most popular material for modelmaking because it can be formulated to achieve a range of properties and it burns out cleanly.

WAXES

Because wax is a familiar material, it's easy to take it for granted. Don't. It can be dangerous. Wax at the dangerous, overheated temperature of 500° F looks the same as wax at 150°F/65°C. Spilled wax can adhere to skin and cause severe burns. Always keep a bucket of cold water handy as first aid for such spills.

Melt wax only in a double boiler arrangement. This can be as crude as a tin can set into a bucket of water, but it is very important. As long as there is water in the larger vessel, the wax won't be heated much above the boiling point of water. If you melt wax directly on a burner or in a flame, there is the possibility that it will reach its flash point and explode. This can spray burning wax great distances and cause serious injury.

Most waxes expand as they melt. Because of this, it is unsafe to set a pot of solid wax onto a burner. The lower section will melt and expand but the still solid cap will seal the pot, allowing pressure to build up. At some point the cap will be weak enough and the pressure great enough to cause the wax to erupt, again with potentially disastrous results. Tip the pot like this, either when allowing the wax to harden in the pot, or when remelting.

Waxes are made of three kinds of ingredients, selected, purified, and blended to achieve specific properties.

1. WAXES - white beeswax, carnuba, candelilla (from palm trees), ceresin, ozenite (earth waxes), and synthetics are all used.

2. RESINS - These are tree saps like damar, balsam, kauri copal (gum), shellac, and rosin from pine trees.

3. FILLERS - Talc, starch, chalk, soapstone, pumice, and wood flour are fillers.

Every wax has its liquidus and solidus points. The former is the temperature at which the wax melts, the latter is the point at which it hardens. In between there is a temperature spread, called the *plastic range*, in which the wax is easily worked. It will be beneficial to know these points and the range between them. A sense of the temperatures can suffice, but for really accurate work it's worth the effort of making a test to determine the exact range of the wax you are using.

Melt a small quantity of wax in a tin can, set in a double boiler arrangement. With a candy thermometer, measure the temperature of the wax when it has melted into a puddle. When all the wax has melted, remove the can from the double boiler. Watch the thermometer and note when a skin forms on the wax. These two points are the liquidus and solidus temperatures.

Carving Waxes

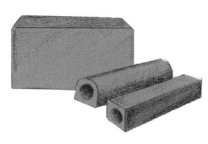

These waxes are worked reductively (also called subtractively). They are whittled or filed away to create a shape, similar to the process of a sculptor working in stone or wood. Carving waxes are available in blocks, tubes, rods, and sheets of varying thickness and are made in several degrees of hardness. Most are too tough to be bent or molded in the fingers. Because each manufacturer uses its own color code, don't assume that a blue wax from one source will behave like a blue wax from another company. Experiment to choose a wax that suits your personal work style.

Blending Waxes

Waxes are purified, refined and blended to create materials with specific toughness, flexibility, viscosity and melting points. They are dyed to color code the various types and to prevent eyestrain when carving. Any wax, from candles to household paraffin, can be used for modelmaking, but most modelmakers prefer to buy a ready-to-use product.

Unique waxes can be created by blending two or three casting waxes. Because these are clean when purchased, mixing them almost always yields a workable wax as long as you take care that the wax is kept clean and is not overheated as it is being prepared. Burning out some of the ingredients will change a wax, usually for the worse, leaving it pale, brittle, and sometimes full of air bubbles.

PREPARING A BLOCK

Any shape can be cut down from a single block, but there are cases where this is time consuming and wasteful.

Often it's more efficient to weld blocks of wax together to create a generalized outline of a design.

To fully bond two blocks of wax, both pieces must be molten at the point of contact. This can be achieved in the flame of an alcohol lamp or with an electric soldering pencil. Press the fluid surfaces together with a slight twisting motion and allow the wax to harden completely (about a minute, or until the original color returns) before beginning to carve. Be careful not to touch the molten wax: it will stick to skin and can cause a nasty burn.

CUTTING

Any sawblade can be used to cut wax, but some are better than others. A jeweler's blade will cut, but the friction heat it generates melts the wax along the kerf, or line just cut. If the blade halts for even a few seconds, the wax hardens (freezes) along the blade and locks it into place. A bandsaw, hacksaw, or coping saw will avoid this problem, but cannot cut around tight corners.

A spiral sawblade is a jewelers' blade that has been twisted during manufacture so its teeth project outward in all directions. This cuts a wide swath that does not seal up on itself. It can be a little difficult to guide with accuracy, so cut outside the intended design.

Spiral blades can be bought from most suppliers of jewelry tools, but if one is needed in a hurry, you can make your own. Grip the blade in a vise at one end and a pair of pliers or a pin vise at the other. Heat it to a bright red with a torch flame and twist. You'll probably find you can keep only about a half inch red at a time. Twist this, then move down the blade to twist the next section. Because this blade will be used only on soft materials, it's not necessary to harden and temper it.

FILING

Most carving of hard wax is done with files. Coarse-toothed files will cut quickly and resist clogging. Those sold for use on soft materials such as wood, plastic, or white metal will be good choices and can be bought at hardware stores or jewelry supply companies. This kind of file is typically large and heavy, which can be awkward on a jewelry scale, especially when working on a lightweight piece of wax.

For more delicate work, jewelry and casting supply companies sell small coarse-toothed files made for wax work. These are not cheap, but they are the right tool for the job. If they are reserved only for wax, as they should be, they will last a lifetime. Any of the files found on the metalsmith's workbench can be used on wax. Fine teeth cut slowly and are likely to clog, but these are matters of inconvenience and can be overcome. A light coating of talc, cornstarch, chalk dust or silicone mold release will help files resist clogging. Use a file card or fine brass or steel brush to clean files as needed, but don't try to burn off wax that is stuck to a file. This will only make it stick harder.

It's especially important to clean files before using them on metal again because small bits of wax can play havoc with soldering. Also, small bits of metal can be transferred from files to waxes, where they become inclusions in a final casting. The ideal arrangement has separate files, burs and work areas for metal and wax.

Burs

Rotary files and burs used in a flexible shaft machine offer a quick and fluid way to carve wax. Again, coarse-toothed tools are preferred. A steel bur that as only three blades and looks like a propeller is made especially for wax carving. These are available in a spherical shape from most jewelry suppliers.

Rotary tools (i.e. flex shaft burs) heat up quickly, and even the best can clog. This is a nuisance and slows down the creative flow of carving. To avoid this, use a stroke that touches the wax with the tool spinning, then lifts up for a second before touching down again. This touch-and-go method allows heat to dissipate without altering the position of the hands. This fluidity of movement is important in achieving continuity of the form.

The enterprising modelmaker should consider the possibility of making or modifying tools as needed. The time spent will be well repaid in efficient carving. Solder on bits of brass, nickel silver or steel to an old mandrel or a nail.

Scraping

A sharp edge held at right angles to the wax can be pulled along to shave off a small curl of wax. This leaves a smooth finish and can offer a lot of control. Scraping is done with a knife blade, a razor, a triangular scraper, or an improvised tool that has been shaped to create the contours needed for a particular job.

Scraping, or any other kind of cutting with a blade, is usually difficult at the early stages of a model. Note that scraping tends to modify an existing shape. It's awkward to scrape a square block into a ring shank, for example. In this case it is better to first use a saw and then a file to achieve the general contours of the piece. Once the shape has been roughed out, scraping can be an effective way to sharpen a design.

Dental tools make terrific scrapers. They can be purchased through jewelry suppliers, and are sometimes available through a dentist, whose broken tools still have great potential for the modelmaker. They are usually made of unhardened stainless steel and can be shaped with files and sandpaper.

To make a radical bend, heat the tool in a torch flame and bend it at red heat. Forging must also be done at red heat, and even then it is hard to control. Special shapes can be made up from brass or steel and silver soldered onto the tool. These tools suffer no wear in regular use, so the time spent in making them will be repaid over many years of use.

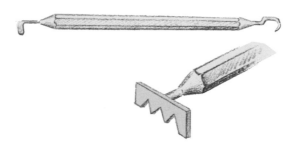

TEXTURE TRANSFERS

Carving wax can pick up textures from any surface that is free of undercuts and able to withstand about 325°F/165°C. Hold a block in a gloved hand or vise grip pliers and warm one surface with a torch flame or an electric pencil. If working "in the field," a cigarette lighter will do. When the wax is uniformly shiny, press the block firmly onto the surface. In most cases, no release agent is needed.

Machine Tools

Carving wax can be shaped on a lathe, with a vertical mill, or in any similar power-er tool. Specific applications will vary with the scale of the project, the tools at hand and the imagination of the modelmaker. Here are a few suggestions:

Small scale spindle turnings can be worked on a drill press or a flexible shaft held in a vise. Drill a hole through a piece of wax, then wrap the drill bit with tissue paper. Slide/twist the wax firmly in place. Run the machine slowly, scraping away wax as it turns. Use a knife blade, gravers, files or scrapers to remove the wax. Wear goggles and a dust mask. The trick here is to work slowly, being careful not to dig too deeply in a single pass. This might cause the wax to slip on the drill bit, making it necessary to refit the bit.

Face Plate Turning

Face plate turning can be improvised as shown. The spindle is a bolt and the plate is a washer. Of course the disk can also be cut from a sheet of brass or nickel silver, or a coin can be used. Be careful to make all sharp edges blunt by filing them.

Heat the wax until its skin is molten, then warm the spindle and press the two pieces together. Small holes in the disk will increase the grip between the two pieces. When cool, tighten the spindle into a drill press or flex shaft and begin turning. Wear goggles and a dust mask to protect against airborne particles.

It is again important to proceed slowly, because too deep a gouge will tear the wax loose. Because it's difficult to recenter the wax on the spindle, caution is more efficient than repair.

Another method uses a preheated soldering pencil to create a cavity of molten wax in a block. Quickly slide a warm bur or nail into the cavity and hold it there while the wax hardens around it. Use a third-hand device to keep the piece steady as the wax cools. Any of these methods, or variations on them, can be used to fix wax in place for milling on other machines.

Gauging the Weight

A typical problem among people learning to use carving wax is the tendency to create heavy pieces. By understanding the reasons for these pitfalls it is possible to avoid them.

Carving wax is brittle, especially compared to metal, the material familiar to most jewelers. After several breaks, beginners become wary of the fragility of wax and hesitate to make it thin. Also, the wax is between ten and twenty times lighter than the metal of which the final piece will be made. A proper model for a jewelry piece will probably weigh so little that it can hardly be felt in the hand. This is a foreign sensation to most metalworkers, and one we are not ready to accept instinctively. It's important to remember that the final object will be made of metal, and will have metal's strength, weight and cost.

Metalsmiths familiar with fabrication should think in terms of familiar points of reference. Is that section as thick as most ring shanks? How thick is this section in B&S gauge? What thickness would you use in this situation if you were fabricating instead of casting?

Thickness can be observed by holding the wax model up to the light, where pale shades will indicate thin sections. As a guide, file a wedge of wax that can be measured and use this as an index of color-to-thickness relationships. A stone gauge or similar measuring device is useful here.

REDUCING THE MASS

First concerns about thickness are taken up in the overall shape and proportions of the piece. If the wax is too heavy when the shape is completed, it is sometimes possible to hollow out sections of the model from the back. This can be done with a variety of tools, but scraping or rotary files are the most frequent choices.

Consult Chapter 2 for information on weighing the model to determine the weight of the finished piece.

Finishing Hard Waxes

After shaping, carved wax is smoothed with fine files, a nylon stocking or a paper towel. These materials are preferred over sandpaper or steel wool which can leave grit or fibers stuck to the wax.

The final surface should duplicate as nearly as possible the finish intended for the metal. Remember that a few light touch-up strokes on the wax will save minutes of work on the metal. Besides taking time, neglecting finishing touches on the wax wastes metal and diminishes details on the piece. Make the wax just right.

Soft abrasives such as a wad of nylon stocking tend to follow a surface, accentuating its characteristics. If a surface is slightly wavy, the wad wears down the hollows and increases the waviness. Soft abrasives are used for rounded, organic shapes.

When the shape calls for a flat facet, use a stiff tool like a file or coarse paper glued to a board (or popsicle stick). For final finishing, use a stiff fabric like denim or twill stretched over a flat stick.

Scrapers can create a subtle linear effect. To make the lines more obvious, file serrations in a scraper blade. These can be used in a crosshatch to make a matte area. This is recommended for areas that will get a lot of wear, that will be difficult to polish, or that will be colored (oxidized). Where a matte area is too confined to allow scraping, use the point of a needle to create a stippled surface.

Repairs

To repair breaks or make design changes in a wax model, heat a pin tool in a lamp or torch flame and touch it to the break. Be certain that all wax surfaces being joined are molten or they will not bond securely. When the pieces have been joined in this way, a bit of molten wax can be dripped onto the heated area to replenish any gap that may have formed. In this way a design or texture can be "erased." Beginning modelmakers should take note of this, and experiment freely as they create a form.

Whenever melting wax, avoid breathing the fumes. They may contain petroleum distillates and can cause sore eyes and nausea.

Scraps

Pieces of carving wax can be recycled as long as they are kept free from contamination. This is important, because even tiny particles of contaminants can become a nuisance if they happen to fall at critical places in a model. Bits of foreign metal can create major havoc in casting, especially if white metals are involved.

Sort through scrap wax pieces and discard any that are unclean. Set the rest in a steel container and warm it slowly in an oven. Take care that the wax doesn't get so hot that it smokes. Heat just until the wax begins to melt, then stabilize the heat at this level. Smelly and noxious fumes may be produced so the room should be ventilated.

It's possible to let the wax cool in this container, but you'll have better results if you pour the wax into a different container. Use heavy gauge aluminum foil to create forms, worked up freehand or formed over cans and boxes. To facilitate release of the wax, coat the foil with liquid soap or silicone spray release.

If you suspect foreign matter in the wax, pour through a piece of wire mesh (something you can throw away later) or a piece of cheesecloth. Allow the wax to cool slowly.

To make sheets of wax, pour the molten mix onto a clean flat surface such as a piece of glass or Plexiglas, or into flat pan that has at least a half inch of boiling water.

Modeling Wax

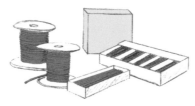

Soft waxes are used to create forms through the build up of successive layers of wax. Both soft and hard waxes are used to create models for jewelry, but the two families of wax are as different, from a designer's point of view, as wood and clay.

Like carving wax, modeling wax is made of a blend of resins, waxes, and fillers. These ingredients are mixed to yield a wax that is soft enough to be bent, cut, molded and pinched into shape.

Modeling wax is sold in bars, sheets and wires of almost any cross-section. It is sold in a rainbow of colors, but as in carving wax, these do not conform to any universal system and are introduced at the manufacturer's whim.

HEAT SOURCES

Heat is essential when working with modeling wax. It is usually provided by an alcohol lamp, but a torch flame or a candle can be substituted. A candle flame colors the wax with soot, and should only be used in a pinch. Whatever heat source you use, reduce eyestrain by arranging a matte black surface behind the flame. This should be a fireproof material such as painted metal.

Lamps

Modelmakers' alcohol lamps are simply glass jars with a tight-fitting metal cap that has a short tube soldered through it. This tube holds a cotton wick, which draws fuel upward by osmosis. A handy feature is a knob on the tube that allows the wick to be raised. As the wick burns it must be extended, and without this feature it is necessary to pull the wick up with tweezers.

An alcohol lamp can be improvised from a small glass jar (e.g. baby food) as shown. An even better solution, because it won't break if dropped, is

made from an oilcan. Saw off most of the spout and drill a small air hole as shown. Over the course of several years this may rust through, in which case it should be discarded and replaced.

FUELS

Denatured alcohol is available from paint stores, where it is sold as a solvent. Lamp fuel is available from department stores where it is sold for use in ornamental lamps. This is often scented, which is unnecessary but will do no harm to the wax. Avoid using substitutes. They are likely to be more volatile than is safe. Under no circumstances use gasoline, or expose the container of fuel to open flame.

It's a good idea to keep a plastic funnel with the alcohol for refilling the lamp. In case of a spill, allow the puddle to evaporate completely before striking a match.

Cutting Sheet Wax

Drafting templates and patterns cut from thin Plexiglas make useful tools. Plastic letter templates can be used with a needle to imprint or cut out letters. Paper punches can also be used to good effect on sheet wax.

Especially in cold weather, wax can splinter when cut. To avoid this problem, dip the wax in a bowl of warm tap water before starting to work. To keep your drawing dry, set a piece of glass or Plexiglas on the drawing.

Models in soft wax grow in the same way a snow drift or an icicle forms, through an evolution of layers. In some cases the first step is to cut out sheet wax with a razor knife or with several passes of a needle point. Because sheet wax is transparent, it can be set directly upon a drawing to copy it.

To type a pattern or message onto sheet wax, warm the wax and shake it dry. Working quickly, sandwich the wax between sheets of plastic wrap and roll it into a typewriter (remember those?). With the typewriter set to "stencil," type the message. To keep the wax soft, position a light directly over the work. With care a hairdryer can be used to keep wax at an appropriately warm temperature.

Tools

Tools in wax working are a direct extension of the hand, and must be comfortable in every sense of the word. Tool length is usually about 6" but this can vary with each person. A basic tool is a needle shape and I prefer a lightweight wooden-handled needle to anything else I've tried. These are available from clay suppliers (sometimes called a cut-off needle) or from a school bookstore where they are called biology needles. I like the lightweight

and I often flip the tool over as I'm working to use the dowel handle to smooth or bend the wax. Other modeling tools can be made from a piece of coathanger, from scraps of brass, or from dental tools. Handles should be insulated with wood or tape.

Modeling tools have a great deal to do with the final character of a piece. Experiment with tools and invent some of your own as you evolve a personal style.

To make a delicate and handy snips for cutting wax wires, grind a thin cutting edge on the tips of a pair of fine-pointed tweezers. The beauty of this tool lies in its light weight, quick action, and in its ability to reach into tight spaces.

Electric Wax Pens

These popular tools offer an alternative to the heat-and-work, back-and-forth motion of modeling with a flame-heated tool. The pen is constantly heated from within, saving time and allowing more concentration on the modeling itself. Inexpensive electric pens must be occasionally switched off (or unplugged) to prevent overheating, but the better quality tools are regulated to maintain a constant temperature. When using these, it's a good idea to experiment to find a temperature that suits the wax being used. Once this is established, it's possible to dial the correct temperature and focus attention on the process of modeling.

Do It Yourself Wax Pen

Soldering irons and wood burning tools can be used for wax work, but they get too hot. Mount a standard outlet into a 2-gang electrical box. In the space beside it, install a dimmer switch to control the heat. Make a short electrical cord with a plug on one end (I use an extention cord with the socket end cut off), and wire this to the outlet.

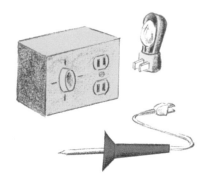

Use a nightlight in the extra socket to indicate when the tool is on. Follow the manufacturer's instructions regarding the wattage that can be passed through the switch. You might want to write numbers on the box so you can calibrate your work. With a little experimentation you'll be able to dial the correct heat for each of your waxes. The total cost, including the soldering iron, was $40.

Most soldering irons make it easy to improvise your own tips. Use a bolt that matches the screw threads on the soldering tip and silver solder pieces of needles, paper clips, or brass wires to them.

Working with Modeling Wax

In some designs it is easiest to start with an armature of wax wires. These are cut to shape and welded by touching the joints with the tip of a hot needle. An armature can be filled in with drips from a wire or with panels of sheet wax. Properly used, an armature not only provides a system for making the model, but a network to assist in filling the mold. It can also be an important structural device to provide strength in the final piece.

METHOD ONE

Hold the tip of a wax wire in the flame of an alcohol lamp. It will melt and form a drop. This can be quickly carried to the sheet and deposited. Timing is critical. If the wire is held for more than a split second, the drop falls off, into the flame. If the wire is not sufficiently heated it will not adhere to the sheet. Practice and patience will soon teach the proper pace.

METHOD TWO

Heat a needle tool and point it at the spot where the wax is to be deposited, keeping it just above the surface of the wax sheet. Touch the wax wire to the tool about a half inch from the tip—the wax will melt, slide down the needle, and fall into place. At least it's supposed to. It is important to control the temperature of the needle.

For large deposits, a wax wire can be laid into position and flooded with additional wax to create a ridge. This wire can be formed from soft wax and can be rolled along the tabletop under the palm to create a "snake" of irregular contours. These can be built up in many layers to quickly establish a thick section.

It is also possible to brush molten wax into place with a paintbrush. Heat a can of wax in a double boiler and use a disposable soft bristle brush.

Sheet wax can be molded in the fingers to make a substructure. Soften the wax by dipping it in warm water or by breathing on it. In very warm weather it might be necessary to "lock" the wax into position by dipping it in cold water. This is especially important when the model will be handled. If handling causes difficulties, the model should be mounted on a stand so it can be worked without distorting it.

A Homemade Extruder

There is almost no limit to the shape of the wires to be created with this tool. Conventional soft wax can be extruded to create wires with unusual cross sections. By mixing a very soft wax, the extruder can be used like a cake decorator's frosting cone to build up interesting shapes.

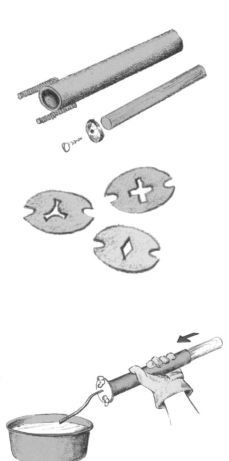

1. Buy a piece of metal pipe about a foot long and an inch in diameter. You will also need a dowel to fit inside this. If a tight fit cannot be found, use a smaller dowel fixed with a rubber washer at the tip to make a tight seal.

2. Cut the heads off two bolts and solder these threaded rods onto the sides of the pipe so about ½" projects below the bottom edge.

3. Use brass sheet (about 18 gauge) to cut out design panels as shown. The space between the lugs must be measured against the two bolts to ensure a close fit.

4. Use heavy gloves when extruding. Pour molten wax into the extruder and slide the piston into place, using slow even pressure to squeeze a rod of wax out the end of the tube. If the wax hardens, work over a small flame or in the warm blast of a hair dryer, rotating the extruder for even heat. Allow the wax to fall into a dish of cool water.

Using a Press Die

Dies may be made of copper or brass through fabrication or etching. These will be the reverse (negative) image of the desired casting. Set the die plate into a vise, supported by sheets of steel and cushioned with a piece of wetted leather or rubber. Warm a sheet of wax and protect it between sheets of plastic wrap. By tightening the jaws of the vise, the die will press into the wax and make an impression. This has many applications for production work.

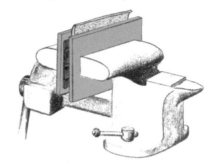

Making Ring Shanks

Warm a steel or aluminum mandrel and smear a layer of wax around it. Before applying the wax, coat the mandrel with a thin film of a parting compound like microfilm, talc, or a spray cooking oil. This technique requires control over the temperature of the rod, and will take some practice, but it's possible to create a band of consistent thickness in a single swipe around the mandrel. Slide this off and invert it on the mandrel to even out the taper. If the wax sticks, cool the rod by holding it under cold tap water for a few minutes before attempting to slide off the wax.

Another way to prepare a ring blank involves building up an armature of sheet or wire. This can be done on a tapered mandrel, but I prefer to work on a dowel with parallel sides. Jewelry suppliers sell a set of stepped aluminum rods with full and half sizes that attach to a stand for a convenient work position.

An alternative is a 6" length of wooden dowel adjusted as needed. To establish the correct ring size, build up layers of masking tape, adding to or removing tape as needed to create a mandrel of the correct ring size. Lubricate the tape to prevent wax patterns from sticking to it, or cover it with a sheet of plastic wrap. Periodically rotate the model as you work to keep it free.

After the basic form has been established, the model is refined by adding or removing wax. Wax can be scraped away with a tool shaped like a miniature spoon made from a brass or nickel silver wire, or from a dental tool. For some waxes (and especially in cold weather) it may be desirable to warm the tool in a flame, but in many cases the wax can be scraped away at room temperature. When warming the tool, avoid using too much heat. Even a slight miscalculation can make a hot tool do a lot of damage. Hold a tissue or rag close at hand and wipe off the excess wax after each stroke.

Remember that heating alone does not make wax go away. It will cool and harden in more or less the same place it was originally. To pull molten wax away, touch it with a hot needle and either allow it to slide down the needle or blot it with a paper towel. To make holes, poke a hot needle  through the wax and pull it back out. Lightly blow at the spot of molten wax and a hole will open up. Droplets will splatter onto whatever surface is below the piece so I suggest setting a piece of scrap paper into position like this.

Finishing

cut here

Files are almost never used with modeling wax. They will rip the form, fill with wax and be totally useless. Sandpaper and steel wool are also inappropriate for modeling wax. Steel wool is especially bad since it will leave fibers of steel that will contaminate the mold and show up in the final casting.

To create a smooth finish on soft wax, heat a needle and draw its tip lightly across the surface of the wax. Some modelmakers slide the model through the lamp flame to smooth its surface, but this is risky. Practice with some scraps before trying it, and then try it only if you are confident of your reflexes. A second too long and your modelmaking efforts will have been wasted.

A pump tip (the kind used to inflate sports equipment) can be modified to make a dollhouse-size Bunsen burner. Attach this to a rubber hose running from a propane or natural gas line and run the flame lightly over the surface of a model to smooth it.

SPONTANEOUS EFFECTS

Some interesting textures and shapes can be created by dripping melted wax into or over a range of materials. Resulting shapes can be worked in any of the methods mentioned above. Never heat wax to its flashpoint! At this temperature (which differs with each wax combination), the wax will suddenly ignite. This is very dangerous. Watch the melting procedure and remove the wax from heat as soon as it is fluid. Always use a double boiler arrangement to melt wax, setting the wax pot into a larger container that has about an inch of water in it. As the water boils away, add more.

Drip wax onto:

- water
  (the temperature changes the effects)

- weathered wood
  (soaked with water)

- a steel slab

- concrete

- leaves

- wood shavings

- plastic wrap

Clay Relief

You can get rich and unexpected results by making a wax impression of forms and textures pressed into clay. Use a potter's clay, or any other clay mixed with water. Plasticine (children's modeling clay) won't work because it melts when hot wax is poured into it.

Roll or pat the clay into a slab and create patterns by (for instance) pressing, dragging, or cutting. Remember

that the results in the wax impression will be the reverse (negative) of the shape in clay.

Pour or brush molten wax onto the clay. When pouring, it might be necessary to build up a wall of clay to contain the pour. When using a brush, "throw" the melted wax onto the clay, using a sharp flick of the wrist to fling the wax into small recesses. Modeling or inlay waxes are preferred for clay relief because they are especially fluid when molten.

The hardening of the wax is usually interpreted by observing a change of color as it cools. After a minute or two, pull the wax from the clay and rinse it off under cool water. A soft brush can be used to clean away the mud that results. Wax sections can be cut with a knife or scissors, and can be bent, formed, modeled and added to as with any other wax model.

ORGANIC MATERIALS

Many organic materials can be burned out and cast. Note that because each material has its unique characteristics, you'll need to experiment. Here are a few of the factors to keep in mind.

• The volume of water in a model will effect its ability to hold its shape during investing and burn out. Plant materials collected in the autumn will be drier than those collected in the spring, and in some cases this can make the difference between the success and failure of a casting.

• "Leafy" materials generally burn out; "woody" materials generally don't. This can be confusing because the difference between these two types of materials is vague. I've found that some pine cones, for instance, burn out cleanly, while others leave a residue of charcoal that prevents the mold cavity from filling. Experimentation is the best way to tell which materials will yield a clean casting.

• The weight and moisture of the plaster-like investment can greatly deform a natural model. A flower in bloom will crumple when covered with investment, so even a complete cast will yield a sorry, bedraggled bud. To strengthen a model against this, coat one side with wax, usually by dripping or painting molten wax carefully into place. A flower could also be sprayed with lacquer or paint. Beware of especially nasty fumes during burnout. Turn on ventilation and leave the room.

• Very thin areas will not cast because surface tension of the molten metal prohibits it from entering a small crevice. Thin sections such as the wings of insects should be thickened with wax, lacquer, paint or glue.

• Bony structures and shells do not usually burn out completely. If you want to try them, build ash traps into the sprue system to collect the unburned residue and use an especially long burn out.

Found Plastic Parts

Most plastics burn out completely and offer a wide range of possibilities for modelmakers. The fumes produced as plastics melt can be toxic, and must be properly vented. Work with plastic models in a large, well-ventilated space. When burning out plastic models, it is especially important to vent the exhaust gases. Failure to observe these safety rules can be risky to your health.

Plastic parts abound in our society: pieces from model kits, machines and household gadgets are all sources for raw material. The huge supply of interesting shapes makes the job of using plastics difficult because there is a temptation simply to transform a plastic article into a metal one. This is a technical exercise, and has little value beyond that. The challenge of found object casting is to make creative use of found objects, giving them a new and innovative meaning.

Hard Plastics

Some plastic items will be too thick to be cast in metal. They can be thinned by filing, sawing, or grinding with burs. Do not use heat, unless you want to distort the object.

It's a good idea to test a sample before devoting much time to a model that might not cast well. If you have several pieces of the found object, it is easiest to test by investing and casting a piece. Any metal can be used for this sample casting.

After casting, examine the work for completeness and surface quality. Missing details or large voids indicate that a residue of the model was left in the mold and prohibited the metal from completely filling the cavity. Perhaps a longer or hotter burnout is needed. Or perhaps that item simply can't be burned out.

For a simpler, less conclusive test, set a piece of the model material in a small can and cover the lid with a scrap of metal (brass, steel, etc). Heat the can in a kiln up to about 1200°F/650°C, then check it to see if there is residue left in the can. If not, there is a good chance the material will burn out cleanly. If there is a gummy residue, there is little hope of getting a good casting. In this case it might be necessary to make a rubber model of the original object and create a wax model from that.

Expanded Plastic

Soft plastics such as styrene and Styrofoam can be used to make models. Refer to the safety warning above, because the fumes from working with these materials can be hazardous.

We are surrounded by many kinds of soft plastics, from food wrapping to packing materials; from coffee cups to building insulation and each of these offers unique possibilities for the modelmaker.

Plastic can be cut with a saw or a knife, can be worked with hot or cold tools, and can be used by itself or in conjunction with waxes. Pieces can be heated in a flame to create rich textures and spontaneous shapes. Remember that because small pieces of Styrofoam are practically weightless, it's easy to get carried away and create models that will be heavy when cast. Experimentation and experience are the best guards against this problem. Pieces of plastic can be assembled with hot wax or with a light glue such as airplane glue or rubber cement. Very frail pieces of plastic can be stiffened by adding a layer of wax or by spraying with a lacquer or plastic spray.

A sprue is a rod that:

- supports the model while the mold is made,
- provides a passageway for melted wax to leave the mold,
- makes the entrance through which molten metal is poured or injected into the mold, and
- positions the model so as to reduce turbulence as the mold is filled.

Sprue Logic

The rules governing sprue placement and size can be reduced to simple common sense principles. By paying attention to the laws of physics and learning from your experience, it's possible to have consistent success in spruing.

SPRUE MATERIALS

Sprues can be made of any wax, but round wire of soft wax is preferred because the suppleness of the wax allows for positioning of the model within the flask. Sometimes you'll have to lean the model to correct its position within the flask and a stiff sprue makes this difficult. Soft wax also has the advantage of making a firm bond when heated. Having a model come loose during investing is a nuisance, and using a soft sticky wax can help avoid it.

Proper casting requires an efficient flow of molten metal into the mold cavity, and this is achieved by keeping the walls of the passage smooth and as straight as possible. A sprue of irregular cross-section will stir up the metal as it enters, so avoid a sprue of wires twisted together. Twisted wires can also cause problems because the thin ridge of investment that forms along the twist can be broken off by the metal as it rushes in, creating inclusions of investment in the cast object. If sprue wire of the correct size is not close at hand, combine wires to make a thicker sprue and roll them together between your palms until

they fuse. Sprues can also be thickened by joining wax rods side by side, or by dipping a wax wire repeatedly into molten wax.

Plastic rods can be used for sprues, but they are not recommended. Because most plastics melt at higher temperatures than wax, a model will be melted before the sprue opening (gate) is cleared. This can allow the wax to boil inside the mold, creating pressure that can crack a mold, and create residue and turbulence that can rough up the cavity wall.

Checklist

1 Determine orientation in the flask.

2 Decide on the number of sprues.

3 Determine the amount of metal needed for the casting.

4 Select a flask and sprue base that will accommodate the model.

5 Mount the sprued model firmly onto the base. Be sure the point of contact has a fillet (a curved joint) rather than a sharp right angle.

6 Invest, wearing a respirator.

Orientation

The model should be placed in the flask so that it has about 3/8" (7mm) clearance on all sides. If several models are cast in the same flask, they should be at least 1/8" apart. The distance between the top of the model and the top of the investment should be somewhat greater, especially in the case of heavy castings where the force of a large amount of molten metal slamming into a thin top wall of investment could cause it to break. A half inch is considered a minimum.

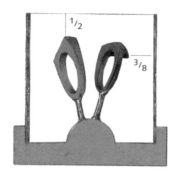

Flat models, or flat areas, should never be horizontal. A tilted model forces the metal to ricochet along the cavity walls, which fills the cavity more completely. If the molten metal is thrown against a flat wall it will be deflected straight back down the sprue, causing turbulence as it bumps into inrushing metal. This can create porous or rough-surfaced castings.

don't trap
air bubbles

The model should be positioned so it is entirely "downstream" from the metal's point of entry. The drawing shows the direction of flow; avoid a position that forces the metal to reverse this direction. In cases where a small amount of metal is involved, especially when it is backed by a relatively larger mass, a small proportion of backflow is acceptable.

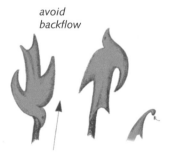

avoid
backflow

flow

backflow in
sections is (

All sprues should radiate outward from the center of the sprue former. They do not have to meet at a central point, but should come together in the top third of the dome. From here they should reach out to attach at each thick area of the model. Refer to the drawings for some typical configurations.

auxiliary
sprue

SPRUE PLACEMENT

1 Attach sprues to the thickest section of the model.

2 Connect sprues where they will not damage surface textures, and where they can be easily removed after casting.

3 Supply each thick section of the model through its own sprue.

4 Thin sections, especially those away from the center of the model, may need separate sprues.

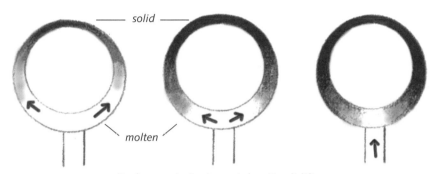

Dark areas indicate metal as it solidifies.

Recrystalization

As metal cools, it falls into an organized pattern in a process called *recrystallization*. In the same way that an orderly stack of bricks takes up less space than a disorganized pile, as the metal cools, or takes up less space. Small colonies or grains of crystals organize themselves independently, creating microscopic voids between them that will remain in the casting as pits unless they are filled. If molten metal is available, it will be drawn by osmosis into these spaces, filling in the voids.

If thicker section will cool a split second later than a thin area and will provide a supply of molten metal to fill in the voids. When this last section is recrystalizing, there is no supply of molten metal to fill in the voids, so it

might end up with pits. The aim of proper casting is to arrange sprues so that this happens outside the cast object, for instance in the button (the lump at the end of the sprue). To achieve this, be sure that the primary sprue is the thickest section within the mold, and the button is the largest section of all. Because it is the largest mass, this lump will be the last part to cool. The series of drawings below show this progressive solidification. In a scale of jewelry, all of this takes place in a few seconds.

See below for notes on the use of reservoirs to direct recrystalization. Related information can also be found in Chapter 10, Foundry Casting.

Choosing a Flask and Sprue Former

The model(s) must have about 3/8" clearance on all sides, with a little more at the top. Using a flask that is too small will risk a thin section in the mold that can break through when molten metal is thrown into the cavity.

A mold that is too large can cause other problems. For one thing, this is wasteful of investment and will require a longer time and more energy to burn out. The gases that occupy the cavity during burnout are normally vented through the pores of the investment and out the top of the mold. If the investment is too thick at the top, this venting action can be impeded and a partial casting can result. Keep the mold between 1/4" and 1/2" thick (6-12 mm) at the top to provide sufficient strength while still allowing the gases to escape.

For special situations, make a mold from a food can. Be sure to use a "real" steel can, not one of the new genera-

tion aluminum cans. Cut off both the top and bottom and, if necessary, cut it down to size. This container can be bent to an unusual shape, as long as it will fit into the casting machine. If you are using gravity, sling, steam, or vacuum casting, this is not a concern.

Because this tin can is an unusual size or shape, standard sprue bases will not fit. Make a base as shown, using either plasticine, soft wax, or earthen clay. Set this on a flat surface such as a piece of Masonite, plastic or a jar lid for ease in moving it around.

ON-SPRUE RESERVOIRS

Some complex sprue systems make is impossible to connect the button directly to the thickest section of a model. In these cases, nodules of wax are set up along the sprue to provide a last-to-cool area outside the model.

Roll a ball of soft wax in the fingers, poke a hole through it with a scribe, and slide it onto a wax wire, then use a hot needle to smooth over the joint between the two. This ensures a nonturbulent flow for the metal as it passes through the reservoir on its way into the mold cavity.

Attaching Sprues

For general jewelry casting, use sprue wax that is slightly heavier than the average section of a piece. I've found 8 gauge B&S to be a convenient size. The wax is soft enough to be easily rolled between the palms into a two or three-ply rod for thick pieces as needed.

Set the model "on its back" or "tummy up" after you have a good mental picture of where the sprues will attach. Hold a needle tool into the flame of an alcohol lamp and position the sprue wire just above the place where it is to join. Touch the hot needle to the tip of the wire and lower the wire into place. Keep the hand holding the wire steady for a few seconds until the wax hardens. While the needle is still hot, cut off the sprue to its correct length.

When the sprues are attached, pick up the model in the finger and examine each of the joints. With the model held in the same position it will have in the flask, check to be sure that each section of the model is downstream from the entrance of the molten metal. If there are any problems, this is the time to correct them.

Turn the model over and hold it in position on top of the sprue former. If necessary, trim the sprues to the correct length, either with a hot needle or by pinching off the wax between fingernails. Get the needle tool especially hot and pass it along the tips of the sprue wires while simultaneously lowering the assembly onto the dome of the sprue base. Repeat the process described above to supply extra wax to create a fillet at the base of each sprue.

Fillets

There should be a fillet, or broadening of the wax at the joint between the sprue and the model. A fillet is a meniscus-curved surplus of wax added to a joint to enlarge the mass and round the form of the joint. Having a straight-sided joint (shown in A) can cause the metal to splatter into the cavity, which is bad. Having a pinched joint (as in B) sprays the metal into the mold, causing turbulence. Also, a sharp edge of investment is fragile and can be broken off by inrushing molten metal. This could result in loss of detail and an investment inclusion.

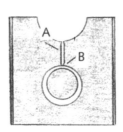

Sprue entrance is restricted: not good.

Rounded openings encourage smooth flow.

Thickening the Point of Attachment

To supply extra wax, heat the needle tool and point it at the joint, keeping the tip about 1/8" away from the sprue. Touch a wax wire to the needle about a half inch above the needle tip and a drop will form, slide down the needle, and fall off onto the base of the sprue. All of this assumes that the needle is the correct temperature. Some practice might be needed to make it happen just like that. This process is repeated for each of the sprues.

An alternate method does away with the needle tool and instead substitutes a quick hand and a sure eye. Having determined the location for a sprue, pass the tip of a wax wire through a flame to create a drop of melted wax at its tip. Quickly set this into position and hold it there until the wax hardens. This technique is a little difficult to master but is very quick once you get the hang of it. Of course an electric wax pen can be used to great advantage when spruing.

MAKING CHOICES

It's almost impossible to satisfy all the requirements of proper spruing in a single piece. Many designs are just too complicated to allow the rules to apply easily. In other cases it's possible to sprue as described, but the resulting structure looks like a wax model caught in the web of a deranged spider. Having to cut away dozens of sprues creates a lot of work and can destroy the freshness of a piece.

Casters who work with rubber molds are in an ideal position to experiment with sprue arrangements. Wax models are easy to duplicate once a rubber mold has been made, so testing identical pieces with various sprue arrangements is relatively simple. Castings that do not completely fill are simply thrown back in the melting pot. Thicker or additional sprues are added sparingly, until the minimum effective sprue system is found.

Even then a risk factor can be built into the casting schedule. If one in ten castings fails, it is sometimes easiest for the caster to invest 10% more models than necessary and play the odds that the result will provide the required number of sound castings.

Thinking Ahead

For beginning casters it's easy to forget that the metal will fill the sprues as well as the model. The sprues will grow out of the casting as solid metal horns, and in some cases these can be difficult to remove. With this in mind, plan ahead by attaching the sprues in locations that can be easily reached with a saw, snips and files.

On models that have an unusual surface texture, the placement of sprues must be cunningly devised so the texture isn't ruined. It is often difficult to duplicate a wax-formed texture in metal after the sprue has been cut away.

On models that have several thick sections, it's wise to run a sprue to each of these. As shown, these can all originate at the sprue base, or can sometimes branch off of one another. Thick sections joined by a relatively heavy bridge can stand alone, especially if the model is fairly small, but when the bridge is thin, a second sprue is needed.

Thin sections at the far reaches of a casting might also need special attention, for different reasons than those mentioned above. Because of its surface tension, molten metal needs an extra push to enter a small space. As a rule of thumb, the higher the melting point of the metal, the greater is its surface tension. Especially at the "end" of a casting (the area furthest away from the gate) the power of the original thrust is diminished. By supplying a separate sprue directly to this area, a supply of metal is fed directly into the tip.

Separate Cases

Avoid damaging the texture.

Sprue each thick section.

Supply directly to extremities.

Some casting methods or materials require slightly different approaches to spruing. These include steam, cuttlefish and sand casting. Consult the appropriate chapters for detailed explanations of these techniques.

Determining the Amount of Metal Needed

Experience brings confidence in guessing at how much metal is needed for a casting. Especially in the case of familiar objects, it's possible to develop a pretty good eye. But for beginners, or in cases where the model is unlike other recent castings, it's a good idea to rely on a system of measurement to determine the amount of metal needed.

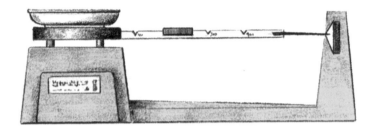

WEIGHT

By calculating the ratio between specific gravity and the relative weights of the model and the metal to be cast, it's possible to figure out how much metal to use. This requires an accurate scale, sensitive to small weights. A penny-weight or gram scale will work well. An inexpensive postal scale is better than nothing, but not much.

Weigh the model with the sprues attached, remembering to pay attention to the units of weight being used. If you start with the weight of the model in grams, for instance, your result will also be in grams. The unit you use doesn't matter; it is consistency between model and metal that is important.

Because the specific gravity of wax is about 1, the math becomes a simple matter of the weight of the model multiplied by the specific gravity of the metal being used. Consult the Appendix for a chart of specific gravity figures for familiar metals.

Be clear in your mind what you have weighted. If you weighted the model and the sprue, remember to add metal for the button, the lump of metal that fills the dome shape of the sprue former. This is important to a good casting since it supplies molten metal and also "closes the door" on the casting. As a rule of thumb for sterling jewelry, add about 10 pennyweights per model for the button. When larger models are involved, the size of the button should grow proportionately.

This system can also be handy for establishing a price or a price differential for a customer. It's possible, for instance, to weigh a model before sprues to determine the amount (and from there the cost) of the metal to be used in it. In this way you can quote on the same object cast in, say, 10k, 14k and 18k gold.

WATER DISPLACEMENT

This method makes an inexpensive alternative for people without a scale. It requires only a glass vessel, like a graduated flask or a clear jar. Put a piece of tape on a jar and add water. Make a mark on the tape to indicate the water level. Submerge the sprued model and make a second mark that indicates the rise in water level. Because wax floats, you'll need to mount the wax on a wire or hold it in thin tweezers to make it submerge.

After removing the model from the water, add bits of metal to the container until the water level has been raised up to the mark. Again remember that this system measures the object that was submerged; you must allow extra for a button. This system is not as exact as the weight method, so I recommend adding some extra metal to be on the safe side.

It's important when using the water displacement method that the container not be too large for the model. If it is too big, the amount of displacement is so slight it cannot be accurately marked.

Submerge the model and mark the rise in water level...

... then add metal to the jar until the water level rises to the mark.

A Simple Balance

A simple scale can be made of any available material as shown in this drawing. It is important that the distances between points be accurate and that the two dishes be of the same weight. These can be hammered out of sheet or be as simple as paper cups. The cross bar can be hung from a string or balanced on a vertical post.

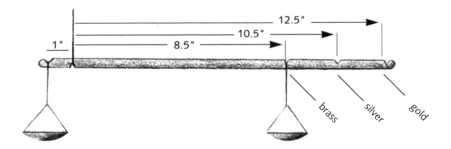

Place the wax model (specific gravity = 1) in the dish on the left. The distances from that dish to the notches in the crossbar represent the specific gravity of the metals shown there. Add metal to the relevant dish and when the bar is level, you have enough metal for the casting. Of course you will want to add more for the sprues and button.

CHAPTER 3

INVESTING

In the early days of lost wax casting, say around 5000 years ago when the Egyptians were doing it, the wax model was dipped in successive layers of fine clay slip. These were hardened and strengthened with bits of fabric or grass to make a tough mold. It was a great idea and has seen few changes over the years. Perhaps the most significant change is the development of a better moldmaking material called investment, a word that comes from the Latin root meaning to *clothe, cloak, or surround.*

INVESTMENT

Investment is a version of plaster that has been formulated to withstand the high temperatures of burnout. It is made up of three basic ingredients:

1. **Gypsum**, a hydrated calcium sulfate. This white mineral, which is the natural material that accounts for the hardening of plaster and Portland cement, makes up 25-40% of the mix. In investment the gypsum is hemihydrated, which means about three-forths of the water in the compound has been removed.

2. **Silica**, a natural white or colorless mineral found as quartz, is purified and ground to a very fine particle size. Its purpose in investment is to cushion the mold during burnout, taking up the expansion and contraction that occurs as the mold goes through temperature changes during burnout and cool down.

3. **Cristobalite** is a form of pure silica, probably caused in nature by the effect of lightning striking certain chemicals. It is manufactured by heating silica to temperatures of around 3000° F. This form of silica was discovered in the late 1920s and has a higher degree of expansion than other silicas, which makes it a superior mold material.

    Besides these three main ingredients, commercial investment contains small amounts of other chemicals (modifiers) designed to contribute to the working properties and strength of the mold. Some of these are wetting agents that increase the ability of the investment to flow over the model, yielding a more complete covering. Other ingredients control the setting time of the investment or help in removing air from the mix. Recently slivers of Fiberglas have been introduced into investments to strengthen them, a technological advance that is reminiscent of the grasses used by the ancient Egyptians.

    Standard investment is used for casting gold, silver, bronze and just about any other metal. The exception is platinum, which requires a special high temperature investment.

BUYING AND HANDLING INVESTMENT

INVESTMENT CONTAINS SILICA, A POWDER THAT CAN DO SERIOUS DAMAGE
TO THE LUNGS. YOU MUST WEAR AN APPROPRIATE DUST MASK WHEN MIXING
INVESTMENT. AN ORDINARY PAPER MASK IS NOT
SUFFICIENT. BUY A MASK SPECIFICALLY RATED FOR
SILICA. SYMPTOMS OF THE DAMAGE DONE BY
BREATHING THIS DUST MAY NOT APPEAR UNTIL
PERMANENT LUNG INJURY HAS OCCURRED.

Because most of the water has been removed from the gypsum in investment, the powder is actively seeking water. For this reason it must be tightly sealed in waterproof containers, such as plastic bags closed with a twist tie. Even at that, the necessary opening and closing of the bag as the investment is used allows moisture from the air to enter and this will in time affect the properties of the stored investment. Buy in quantities that can be used in six months or less.

When using a large drum of investment, it's a good idea to keep a smaller container that can be filled periodically from the larger one. This not only limits the time the larger supply is exposed to air, it also makes it possible to discard this smaller amount without having ruined the whole supply if moisture does get into the smaller container.

Never allow water to get into the investment container. Always dry your hands thoroughly before reaching into the container and locate the storage far enough away from sinks so that no water can splash in. A droplet of water will form a pebble of hardened investment, and continue to hydrate the adjacent powder. If these pebbles lodge in the wrong place in a mold, they can completely ruin a casting. It is for this reason that some people sift their investment, but a better solution is to see that pebbles do not form in the first place.

Investment is sold through most jewelry suppliers and can often be tracked down through commercial casters, depending on what part of the country you live in. For small quantities, up to 25 pounds, it's probably easiest to buy from your regular supplier of casting products. In quantities larger than that, the cost of shipping becomes an important factor. Contact your supplier or a manufacturer of investments to inquire about a regional distributor. It might make sense for several jewelers to cooperate on the cost of a shipment or a trip to the warehouse. It is sometimes possible to find a local large-quantity user and graft an order onto theirs. In a pinch it is also possible to buy investment through a dental supply lab, but I've found these prices are high.

GOALS

There are several ways to mix investment, but they all lead to the same goals.

- proper ratio between powder and water.
- thorough, uniform mixing.
- a bubble-free mix.
- the optimum time interval.

Failure in any of these areas affects mold strength and surface quality. If there are several deficiencies, these flaws accumulate and can result in an unusable casting. Prepare the work area by seeing that it is clean and free of clutter. Timing is important to proper investing so the process must be efficient. Be certain that all materials are close at hand before getting started.

Before Mixing Investment

- Examine the model to be sure it is clean.
- Remove all dry investment from the flask, sprue base and the mixing container.
- Check the fit between the flask and the base to be sure it is watertight. If it isn't, use clay or wax to make a seal.
- Wiggle the model lightly or hold it under cold running water to be certain it is securely sprued and anchored to the base.
- Paint the model with debubblizer or a similar wetting agent.

Investing is Messy

Precautions should be taken to contain the powder and mix, but there will always be some mess to clean up after the mold is made. Locate your casting area away from "clean" sections of the studio and always plan some cleanup time into the process.

Never pour investment down the drain, where it will harden in the pipes and clog them. Investing equipment such as bowls, spatulas, and sometimes even the work surface should be made of rubber. After the investment has hardened, you can flex the rubber to crack off the investment and throw it in the wastebasket. You can also use plastic containers, but they tend to crack after a few uses.

Hydration

Hydrated investment, i.e. investment that has absorbed moisture, can be identified by several clues. Its set-up time will be noticeably extended. When it does set up, the mold will be weaker than a proper mold. This might show up in the form of fins (called *flashing*) on your castings or in severe cases, the mold might break through when the metal is thrown into it. Hydrated molds will also create roughness on the surface of castings.

A question of hydration can be resolved if you have an accurate scale and a sample of the same brand of investment that you know is fresh. Weigh identical volumes of the two samples. Hydrated investment will weigh more because of the water it has picked up. If the sample is more than 20% heavier than the fresh investment, the old powder should be discarded.

An alternate method is to mix a sample of fresh investment and compare the set-up time of this to the set-up time of the questionable batch. If the old investment takes a couple minutes longer to harden, it should be discarded.

The eagerness of investment to find moisture is what makes it uncomfortable on your skin. Gypsum is strongly basic, and it burns the skin in the same way that lye does. Of course investment is not as potent as lye, and for most people a small occasional exposure does little more than feel uncomfortable. It's usually a good idea to supply extra moisture through hand lotions. The base can be counteracted by washing with vinegar after mixing investment, followed with a soap and water wash. If you find that your skin is especially sensitive, apply a generous film of hand lotion before you mix the investment and/or wear rubber gloves.

GLOSSING OFF

A double check on the timing of your investment can be made by watching for the phenomenon called *glossing off*. Observe the flask after the investment has been mixed and poured into it, where close inspection will show a thin film of water on the top of the flask. There will come an instant when this is absorbed, especially noticeable at the top edge of the flask. The effect is like a sponge or paper towel picking up a drop of water. This will happen between 1 1/2 to 2 minutes after the working time of the investment has passed, or about eleven minutes from the time when the investment powder was first sprinkled into the water. If there is a great discrepancy between these figures and your results, check your measuring devices and the temperature of the water.

You'll notice that the setting of investment is an exothermic reaction. Because the process gives off heat, the flask will feel warm to the touch for about the next half hour.

Mixing TIme

Investment mixing has something in common with making a soft boiled egg: timing is critical. The description that follows sounds complicated, and might seem so for the first several tries, but continued experience will establish a comfortable pattern that eventually replaces clock watching with a pace that is intuitive. I think that holds for eggs, too.

Unless specifically noted by the manufacturer, investments have a working time of 9 1/2 minutes, give or take 30 seconds. Proper investing uses all this time, but no more. If the mixing takes longer than the specified time, the investment will begin to harden (set up) before it has settled into place around the model. Continued work with the investment at this stage weakens it, and can result in flashing on the casting caused by tiny crevices or cracks in the mold.

If you complete the investing in less than the required time, air in the investment has a chance to separate out from the water which will leave raised scratchy trails along the surface of the casting as it bubbles upward.

Watching the clock is always a good habit when investing, and especially important when the process is new to you. Think ahead to figure where you should be halfway through the allotted time. By watching the clock you'll know if you are working too fast or too slowly so you can adjust your pace as needed. See the chart in the Appendix for a visual presentation of this. A kitchen or darkroom timer near the investing table will make this clock-watching especially easy.

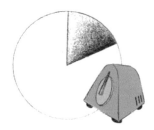

DEBUBBLIZER

Commercial debubblizers reduce surface tension on a model and aid in achieving a complete coating of investment. Its use seems to be important to some casters and superfluous to others. You should try using debubblizer in your work to see if it makes a difference.

Paint the model with debubblizer, using a soft paint brush such as a water-color or calligraphy brush. Allow the solution to dry, which should take about 10 minutes. This is important because moisture on the surface of a model will trap air bubbles and result in a "warty" casting.

Alcohol can be used as a substitute for debubblizer—the same alcohol used as lamp fuel will do, and can be applied with a brush or by using the wick of the lamp.

Water

Consistency, in moldmaking anyway, is a virtue. By using the same water temperature each time, a variable is eliminated from the process, and that is usually beneficial. When problems occur they are easier to track down and correct if there are only a couple of variables.

In most circumstances regular tap water, drawn at room temperature, will make a good mold. Hot water will shorten the setting time of the investment. Room temperature water (72°F/22°C) should feel like dishwater—comfortable, neither hot nor cold. For the really meticulous I'd recommend a flow temperature indicator, a small device attached to a faucet that displays the temperature of the water passing through it. These are available at hardware stores and companies that supply hair salons and darkroom equipment.

In some geographic areas chemicals in the water can affect moldmaking. If you are experiencing casting problems that cannot be explained any other way, try using distilled or bottled water. Some casters have found that letting tap water sit in a loosely-capped jar for 24 hours before using will allow gases in the water to escape and improve results.

INVESTMENT PROPORTIONS

- ● BY WEIGHT

Mix investment in a ratio of 38-40 parts of water to every 100 parts of powder, by weight. Add the powder to the water when mixing as described below. To mix investment "by the numbers" it's important to use an exact scale. As mentioned, the investment must be free of moisture. If the investment has picked up moisture from the air (and remember, it is hungry to do this) it will throw off the calculations.

Refer to the chart in the Appendix to determine to correct amounts of water and powder for the size flask you will be using.

- ● BY FEEL

A less scientific method of determining the water/powder proportions depends on feel and experience. Begin by filling the mixing bowl with enough water to fill the flask two-thirds full. Note the time, then begin to sprinkle powder into the water slowly. After a couple scoopfuls of investment have been added, an island will appear in the middle of the bowl. Continue to add sparingly until the island covers most of the surface and has stopped sinking into the liquid.

Mixing

A strong mold depends on a uniform and complete mixing of the powder and water. The first step in achieving this is to add the powder to the water, not the reverse. If the water is poured over the powder, it forms clumps and cannot be evenly distributed throughout the investment.

Mixing time, even for a small bowl of investment should be at least four minutes. Commercial firms rely on sophisticated equipment that use vacuum or electrical devices or a combination of both to stir the mixture, but for the handcrafter a regular eggbeater can be substituted.

It is also possible to stir the mix with a spoon or spatula, or to knead it with the fingers. I like to use my fingers because I like to be able to feel for lumps as I mix, but I have to admit that a spatula will mix more thoroughly. Perhaps the best arrangement is to start with your fingers, to feel out any lumps, then progress to a spatula for thorough stirring. As you mix the investment, try to avoid creating a froth of bubbles.

Despite the unscientific nature of it, I determine the correct mix by feel. I drag a finger across the bowl, looking for a consistency that feels fluid, but is thick enough to create resistance as the finger is pulled through the mix. The investment should feel a little thinner than sour cream. (It may be necessary to grab a handful of sour cream in order to determine what this feels like). You can read the opaqueness of the solution by dipping a finger into the mix and withdrawing it straight up. If flesh color shows through the coating on your finger, the mix is too thin.

Experience and observation will teach you what to look for and how to know if the proportions are correct. For the first several mixes, follow the procedure described for glossing off.

REMOVING BUBBLES

Even in careful mixing, the investment will be full of air bubbles. Given half a chance, these will attach themselves to model and when the wax has burned out and left a cavity, they will exist as small spherical cavities in the mold wall. When the molten metal enters the mold it will fill all cavities with the result that what were once air bubbles (i.e. places where investment wasn't) become metal nodules or warts on the surface of the casting.

Where there are a couple bubbles on an exposed area of a casting they can be easily removed. They can exist in the hundreds, however, and have a habit of forming in the most inaccessible areas of an object.

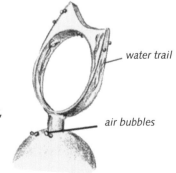

water trail

air bubbles

Vacuum Investing

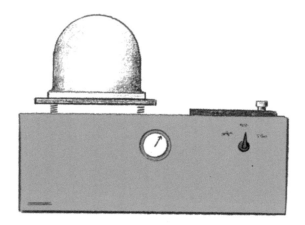

The best method so far developed for removing bubbles is to set the mixed investment under a bell jar and surround it with a vacuum. This pulls air out of the mixture.

If a vacuum machine is available, it should be prepared before mixing the water and powder. Be sure that the table surface and the rim of the bell jar are clean and that an air-tight seal can be made. Mix the investment as described above and set the bowl in the center of the table. Direct the suction to the investment table and turn on the machine. The mix should be vacuumed for about one and a half minutes, during which it will swell in the bowl and become frothy on top, looking like a vanilla milkshake. When the vacuum is released, the mix will fall back to its original volume (or a little less, because it has lost air). The slurry will be noticeably creamier.

To protect the model from shifting under the weight of the mix, pour the investment down the side of the flask, remembering to hold the flask and base together, unless you want a shoeful of investment. The flask, after all, is just a piece of pipe that is lightly set onto the base.

Set the invested flask back onto the table and repeat the process of vacuuming, again for about 1 1/2 minutes. Most machines will hold three large or four small flasks so it's possible to work on several flasks at once.

Don't leave the flasks in the vacuum too long. Prolonged exposure will create a boiling action that can create bubbles in the already creamy investment. To break the surface tension on the top of the mix, rap the spring-mounted vacuum table sharply several times or tap the side of the flask with a screwdriver, hammer handle, etc.

PLAN AHEAD

You will save yourself some cleanup if you plan ahead for the rising of the investment in the flask. As it expands during vacuuming, the investment will spill onto the table and deplete the amount in the flask. One solution is to build up the flask with a collar of rubber or plastic which can be taped in place or held with a rubber band. Ordinary paper can be used, and wide packaging tape is also recommended. Milk jugs can be cut up as a source for the plastic.

For an alternate solution, leave a margin of about 1/2" of exposed flask when filling. After vacuuming, the flask should be "topped off" or "capped" with additional investment. It's all right if this addition has a few more bubbles than the investment lower in the flask because this upper portion of the mold is not in contact with the model, so these bubbles present no problem.

Set the flask aside where it can rest undisturbed for at least 15 minutes but first remember to mark the flask because at this point, they all look alike. Use chalk to write important information (name, date, object, amount of metal) on the flask, and come back later to transfer this information by scratching it into the top of the mold.

Vibration Method

Another way to remove air bubbles from investment is by jiggling them out. The process follows the description above except that the bowl and flask are each, in their turn, set on a vibrating table instead of under a bell jar.

Drugstores sell a body massager that is very similar to an investment vibrator and costs quite a bit less. For practical use it will require some kind of stand to support it. Another alternative is to rig up a small motor with an eccentric wheel. This can be built as a permanent arrangement or to accommodate an electric hand drill.

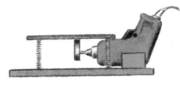

a disk of wood mounted eccentrically on a bolt.

IMMERSION METHOD

In this approach, the flask is set onto a flat surface, preferably a sheet of glass or Plexiglas, and sealed there with clay or wax. Mix investment as described above, then tap the bowl of investment with a rod (e.g., a screwdriver) to loosen bubbles. Pour the creamy investment into the flask until it almost reaches the top and repeat the tapping either on the flask (gently) or on the base. The latter is especially beneficial if the base is set on a flexible surface such as a pillow or piece of foam rubber.

The sprued model is then lowered into the flask with a gentle side to side swaying motion. After the investment has hardened, break the base free and carve a gate in the investment to make a funnel through which the molten metal can enter. This method is often used in conjunction with the hard core method.

HARD CORE (OR SHELL) METHOD

Mix a small amount if investment, about 1/4 cup for a jewelry-size model and paint it onto the model with a soft brush, taking care to dab the investment into nooks and crannies. Sprinkle the painted model with dry investment powder to stiffen this shell. The process can be repeated again, but this is optional and generally considered unnecessary. This layer determines the surface quality of the casting, and is therefore very important. The hardened shell is then given support by surrounding it with investment in any of the methods described above. Because the surrounding investment will not touch the model, it is not critical that it be free of bubbles.

CHAPTER 4

BURNOUT & MELTING

The Purpose of Burnout

- evacuate the mold, (i.e., remove the wax model).
- harden the mold by curing it.
- heat the mold to proper casting temperature.

WHEN TO BEGIN

The mold must be dry before burnout begins. For a small flask (say 2"x 2") this will be at least two hours; for a larger flask, allow a minimum of three hours. The drying can be hastened somewhat by setting the flask in a warm place like on a radiator, in the sun, or in a kiln turned on to its lowest setting.

If a damp mold is heated too rapidly, the water in it will turn to steam which can shatter the mold as it expands. When hurrying a mold, check it frequently during the first 30 minutes of burnout. If steam is rising from the flask, or if there is a film of water on the investment, remove the flask from the kiln and allow it to dry further before resuming burnout.

Steam De-Waxing

Some caster use steam to remove most of the wax before burnout. This reduces the nasty fumes associated with burnout, promotes a cleaner surface on the castings, and restores moisture to the outer surfaces of the flask that dried sooner than the inner area. This uniformity can prevent cracking as the mold is heated later.

Commercial units are available for large shops, but small shops can make do with a simple de-waxer like this. Set invested flasks sprue down onto a screen, over a pan that is three-forths full of water. Set the pan on a hot plate and keep it at a low boil. The steam will penetrate the mold and heat the wax sufficiently to make it drip out. Allow about an hour, after which the flasks can go directly into a warm kiln to complete the burn out. This process is more effective with modeling (soft) waxes than with carving waxes.

Equipment

In industrial situations, a gas-fired baffle kiln is preferred for burnout because of the relatively low cost and great control it provides, but most jewelry situations use small electric kilns. The size of these depends on the volume of work being done—even in a large shop a medium-size kiln is generally sufficient. For small shops an enameling kiln is typical. Flasks can be stacked as long as about an inch of clearance is left from the walls.

Though perhaps forbidding at first glance, kilns are really quite simple and can easily be repaired or improvised. See the following pages for further information on this. ALWAYS TURN OFF A KILN BEFORE CHECKING IT. Failure to do this will provide a quick introduction to the Toaster Effect. Remember when your Mom told you not to reach into the toaster with a knife because you'd get a shock? Well, the same principle is at work here, only on a grander scale.

ADDITIONAL EQUIPMENT FOR BURNOUT

- long tongs.
- thick work gloves (not asbestos, for health reasons).
- a mesh of nichrome or similar high temperature wire.
- a ventilation system (see below).
- a flask hook (make it yourself from welding rod).

FUME CONTROL

The fumes produced during burn out are noxious. They don't smell good or taste good, and they're not good for you. For the sake of your health, install an active ventilation system; one in which driven air pushes fumes outside.

For a clever way to limit the volume of the nasty fumes, use a tray and a matching support either purchased or improvised. Because this wax removal happens at low temperatures, a catch tray can be made of any convenient sheet metal. The idea is to heat the flasks just hot enough to melt (as opposed to burn) the wax, which, depending on the wax will be between 200°–300°F/100–150°C. Stabilize the kiln at the temperature at which the wax first starts to melt and drip out of the mold. Remove the tray when the dripping stops. The wax can be poured into a can to harden and be discarded as solid waste. Because some of its ingredients have been removed, I don't recommend trying to reuse it. After a large volume of the wax has been removed in this way, heat the kiln to the high temperatures necessary to vaporize any remaining residue and clear the pores of the investment.

Preparation

When the investment has dried, pull off the sprue base with a twisting yank which will break the sprues free from the dome-shaped sprue former. Check this area in the mold to be sure the sprue openings (gates) are unimpeded. If there is any question, use a knife to carve away investment. Wash away chips of plaster under running water, using a toothbrush to ensure that the area is free of debris.

This is safe to do now because the mold is filled with wax. After burnout this would not be wise because debris might fall into the cavity and become an obstruction. Working over a wastebasket, use a knife to scrape spilled investment from the sides of the flask. Buildup on the flasks will cause uneven heating and can make it difficult to seat the flask in the casting machine.

If the vacuum assist casting method is to be used, carve a slight recess around the top of the flask, say an eighth of an inch or so. This creates a vacuum chamber along what will be the lower edge of the flask during casting. The pull of the vacuum can be extended into the flask by making channels along the flask wall, made by setting wax rods into position before investing, by setting pieces of coat hanger into place before investing, or by drilling after the investment is hard. If drilling, reserve an old bit for this task because it is likely to rust.

It's also important when vacuum casting that the top edge of the flask make a good seal. To check this (before burnout, when the flask is easiest to handle) run an old file across the top rim. Low spots will show up as dark— continue filing until they disappear.

Position

Flasks are set into the kiln with their openings down and should be raised off the floor of the kiln slightly to allow for the venting of gases. Use enameling mesh, pieces of dry investment, pumice chunks, nails, or pieces of coat hanger wire to raise or tilt the flask.

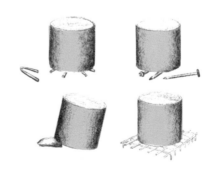

Burnout Procedure

As with many other things in life, burnout can be done the right way or the fast way. The right way uses a schedule of slow heating to guarantee the safety of the mold and the success of the casting. The fast way is usually adequate, but always involves a small risk.

THE FAST WAY

Set the dry flask into a kiln at room temperature and turn it on. If there is a regulating switch, set it to medium. If there is only one temperature, prop the door open. When there is a clear odor from the melting wax, or when you can see the wax dripping out of the flask, turn the kiln up to high (or close the door). Don't breathe the fumes. If possible, leave the room.

Check the kiln every half hour or so until the burnout is complete. This will be indicated by the disappearance of sooty black stain around the mouth of the cavity (the gate). When the stain is completely gone, the residue has been vaporized and the mold is ready for casting. Do not overheat. At the correct temperature, the flask is a dull red in dim light. It should never be heated to a glowing cherry red.

Allow the flask to cool down slowly to a temperature that is about 300°–400°F/150°–200°C less than the melting point of the metal being cast. This can be done by leaving the flask in the kiln, now turned off, with the door open. Remember that at this stage the mold is fragile and should be handled with care. Dropping or bumping it can break thin investment dividers within the cavity.

THE RIGHT WAY

As mentioned before, the value of consistency is that it makes it easy to repeat success. By using a scientifically tested burnout schedule it's possible to guarantee results time after time.

---

3 HOUR CYCLE

for flasks up to 2" x 2"

Hold time	°F	°C
1 hour	300	150
1.5 hours	700	370
0.5 hour	1250	675

---

5 HOUR CYCLE

for flasks up to 3" x 3.5"

Hold time	°F	°C
2 hours	300	150
1.5 hours	700	370
1 hour	1000	540
0.5 hour	1250	675

Kilns

Basic components of a kiln are firebricks for the walls, floor, and ceiling, a shell to hold the bricks together, and a heat source. For heat in an electric kiln, add a coil of resistance wire, a switch and a cord. For heat in a gas kiln use a torch (burner) and a fuel source.

Soft firebricks are available from a ceramic supply company for a couple dollars apiece. They can be cut with a hacksaw and drilled with a conventional bit. Joining is done by cementing with refractory mortar or with corrugated fasteners.

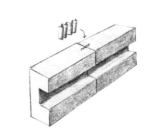

The heating elements in an electric kiln are made of nichrome or a similar alloy and rely on a specific ratio between the diameter of the coil, the thickness of the wire, and the total length of the coil to achieve the correct temperature. If you are repairing a kiln, contact the manufacturer and include the model and serial number to buy the right replacement. If you are building a kiln from scratch, figure out how long the coil will be by measuring the track it is to follow (use string). Send this information, along with temperatures you need to reach, to a ceramics supply company and ask their advice.

The shell can be elaborate or so plain as to be almost non-existent. A sturdy shell will increase the life of a kiln by many years, but if your needs are immediate, you can get by with some steel wire and Scout knots. For a small kiln, an acetylene/atmosphere can torch with a large tip can be used as a burner. For a larger kiln, you can buy a burner from a supplier of ceramic equipment. The drawing indicates a way in which the flame can be diverted through the bricks to create a uniform heat without blowing directly on a flask.

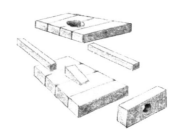

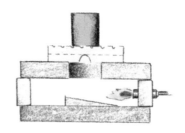

VENTILATION

THE FUMES PRODUCED BY BURNING WAX ARE UNPLEASANT AND UNHEALTHY. THEY MUST BE VENTED! THE SCALE AND SOPHISTICATION OF YOUR VENTILATION SYSTEM DEPENDS ON THE FREQUENCY OF YOUR BURNOUTS, THE AMOUNT OF WAX BEING VAPORIZED AND THE SIZE OF THE ROOM IN WHICH YOU WORK.

The best ventilation system is motor-driven and uses suction to send fumes directly outdoors. It's possible to attach a vent to the kiln, so no gases enter the room at all. This is the most complicated system to install and it will require extra energy for burnout because heat is being constantly carried away.

A hood-type exhaust is the next best approach. A kitchen range hood fitted over a window directly above the kiln will remove most of the fumes.

An open window is pleasant but doesn't count as ventilation unless there is a forced current of air pushing the fumes out. In certain low-volume situations an open window in conjunction with a household fan will suffice.

In lieu of these, or even better, along with these systems, do the burnout in a room that can be ventilated but closed off from the rest of the work area while the fumes are present.

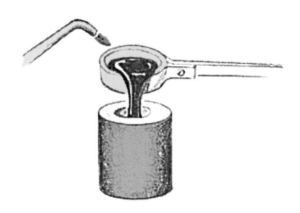

DO NOT OVERHEAT

Wax vaporizes at 1250°F (675°C).

There is never any reason to go above this with wax models, and rarely does it help with nonwax models. At temperatures above 1300°F (700°C) the gypsum binder of the investment, calcium sulfate, becomes unstable and can break down into its components. The sulfur thus released lays in wait in the mold cavity, eager to attack and stain metals, especially silver. Not only will the resulting casting be dark, but the sulfide layer can be so deeply and firmly bonded to the metal that it cannot be removed. This breakdown can also be accompanied by a loss of detail in the mold wall.

CHECKING THE BURNOUT

To determine when the burnout is complete, make a visual inspection of the mouth of the flask. First, turn off the kiln, then grab a flask with tongs and pull it out, inverting it so you can see the gate. If there are black stains in the investment, there is a wax residue throughout the mold and it is not ready to cast. Return the flask to the kiln and continue the burnout.

If the investment around the mouth of the mold is all white, or shows only a ghost of the former stain, you can be assured that the wax has been vaporized and the mold is ready to be cast.

The Physics of Melting

PROPER MOTOR-DRIVEN VENTILATION SHOULD BE USED DURING THE MELTING OF ANY METALS IN ORDER TO ELIMINATE FUMES THAT MAY CONSTITUTE A HEALTH HAZARD.

Most metals at room temperature are held together by the electromagnetic bonds between their crystals. The molecules and atoms that in turn make crystals are held together by electric and nuclear forces. When sufficient energy (in the form of heat) is supplied to a metal, the bonds holding it together undergo a readjustment, replacing a lattice arrangement with a random one. A flat sheet of metal melts and is drawn up into a blob.

As the crystals of metal move about, they are especially eager to combine with oxygen, which forms compounds called oxides. When melting metal for casting, it's important to prevent this from happening because oxides cause pits and stains in the casting.

One way to reduce the formation of oxides is to limit the volume of oxygen in the area around the molten metal. Some torches and furnaces supply a jet of an inert gas such as nitrogen that surrounds the metal and seals it off from oxygen contamination.

Another approach uses a chemical screen against oxygen. The chemical, called a flux, covers the metal with a film that absorbs oxygen rather than surrender it to the metal. See below for a list of fluxes.

Another way to minimize the formation of oxides, along with flux, is to provide a heat source that is reducing, or fuel-rich. This kind of flame has more fuel than can be consumed by the oxygen around it and keeps oxygen away from the metal by forming fuel/oxygen compounds.

Torch Melting

Neutral

Oxidizing

Reducing

Any torch that provides enough heat to completely melt the metal can be used for casting. As the charge, or amount of metal being used increases, the torch size must also increase. The melting point of the metal is constant, but more energy (e.g., a bigger flame) is needed to complete the melt. Heating will be most efficient if the metal is in the form of small pieces rather than a large lump. The greater the surface area, the faster the melt, and this is best achieved with many small pieces.

For small amounts, say under an ounce of metal, an acetylene/atmosphere torch can be used. For larger charges, (and always for platinum) a torch that combines fuel with oxygen must be used. Propane or natural gas are preferred over acetylene in this arrangement because they are cleaner fuels. Acetylene can affect the crystal integrity of the molecular bond in the casting.

Use a large torch tip and adjust the fuel/oxygen proportions for a reducing flame. This is characterized by a large bushy flame, a rich blue color and a low roar. Its opposite number, an oxidizing flame, is evidenced by a pale lavender color, a sharp pointy flame, and a hissing sound.

All torches require a supply of air around the tip to work effectively. Do not push the tip into an enclosed space, such as the mouth of a crucible because the flame will go out. Instead, crank up the flame so it can do its job from a couple inches away.

Some torch manufacturers offer a special tip for melting that has a large nozzle and is usually about a foot long to keep the operator safely away from the heat of the melt. Ask your supplier of torch equipment for a recommendation about which tip to use. The tip will be supplied with recommended pressures and you should use these numbers when setting your regulator to provide the most efficient flame.

FURNACE MELTING

It is common, especially for gravity or vacuum-assist casting, to melt the metal in a kiln or furnace. For large charges, this a little easier than holding the torch because the operator does not have to be physically involved for the whole time of the melt.

It's possible to use the same furnace for burnout and melting, but it is rarely efficient. The needs of the two pieces of equipment are different, including different temperature ranges, venting, access and size. A melting furnace should be just large enough to hold the flask, should heat up quickly to temperatures around 2000°F /1100°C, and should be sealed tightly enough to restrict the infusion of air. These are all opposites of the well-vented, slow-heating needs of burnout.

Metal is typically melted in a cylindrical crucible, usually made of graphite or silicon. When the metal is molten, the crucible is picked up with tongs and metal is poured from it into the mold. The furnace and the mold should be side by side, or at least no more than a couple steps apart.

A popular alternative is the self-contained melting furnace which consists of a cylindrical jacket of firebrick or insulating material laced with heating coils. A crucible is slid into place and the unit is plugged into a wall outlet. When the charge is melted, the whole unit is picked up like a tankard and the metal is poured into the mold. Small models have the heat control in a separate box, while larger models build it into the base of the cylinder. The fanciest versions of these furnaces are equipped with pyrometers and thermostats.

Two Melting or Burnout Kilns

A melting furnace such as this can be built from fire bricks to work with a large annealing torch. For charges of less than 10 ounces of silver (7 ounces of 14k gold), this arrangement is very efficient. For larger melts it might be necessary to make the larger furnace shown below.

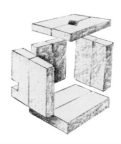

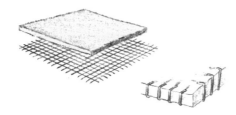

Grip a sheet of Fiberfrax aganist a piece of hard-warecloth by bending over "fingers" from the edge of the mesh.

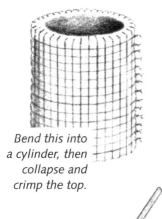

Bend this into a cylinder, then collapse and crimp the top.

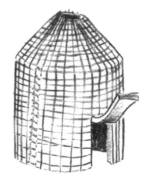

Cut a door flap and position a burner head or large torch in such a way that the flame enters there.

Fluxes

Flux is sprinkled onto the charge as soon as it is warm and again as it becomes fluid. A pinch is usually enough for most jewelry-size castings; too much flux can result in flux inclusions in the casting and can also shorten the life of the crucible.

Most suppliers of casting equipment sell a commercial flux preparation, of which borax is a basic ingredient. Plain powdered borax can be used as a flux all by itself: it is available through jewelry suppliers, chemical companies, drugstores and some supermarkets.

Ammonium chloride (also called sal ammoniac) is an effective flux for gold. It leaves the metal clean and contributes to its working properties because it provides for a better recrystalization as the metal cools. Ammonium chloride produces poisonous fumes and should only be used in small quantities and where active ventilation is in use. Wear a respirator.

OVERHEATING METALS

Yes, it is possible to overheat, or "boil" metals. The effects of this will vary depending on the metal, but they are always bad. Some alloys will separate, some will take on unexpected characteristics such as brittleness or staining, and some will change color. Almost all will absorb oxygen. The results in the casting will be pits, coarseness, brittleness, and staining.

When metal is completely fluid it will lose any former shape and conform to the shape of the crucible in which it sits. In most crucibles, this means the metal is a smooth-edged oval lump. Jiggle the crucible to swirl the molten metal and reveal any solid lumps in the otherwise molten mass.

When the metal is heated above its melting point, its surface starts to move in a swirl all by itself, which indicates the correct temperature for casting. At this point the metal is sufficiently hot to remain molten for several seconds after the heat source is removed, but it has not yet been heated to a damaging temperature.

If further heat is applied, the swirling motion gets faster and the metal makes tiny jerks. This is called "boiling" and is the thing to avoid. If you have accidentally heated to this point and realize it right away, probably little or no harm will be done. Remove the heat and add flux. Do not let the metal cool down to the point where it will harden, but allow it to cool to the proper pouring temperature.

CHAPTER 5

THROWING THE CASTING

Gravity

For metals that flow very well (usually these have a low melting point), gravity is an ideal force to fill a mold—it's free and always available. When using gravity, it is especially important that the model be oriented in the mold to avoid backflow. To ensure that the mold is completely filled, it is sometimes necessary to design a sprue system that includes sprue-like forms called risers and vents. Detailed information on these is given in Chapter 10, Foundry Casting.

CENTRIFUGAL

Centrifugal force is the most common technique in jewelry casting. This push is necessary because the strong surface tension of jewelry metals inclines them to form a ball when melted. Gravity alone can be used for silver and gold, but there is a chance that the metal will curl away from the mold, resulting in a loss of detail. In the case of an object with thin lacy areas, the metal will refuse to enter thin parts of the mold altogether.

Force is needed to fill a delicate mold, but great force can be a detriment. The idea is to have a constant and even pressure driving the metal into the cavity. If the mold is delicate, the force of the metal slamming into the mold can break off tiny bits of investment with a resulting loss of detail. Also, if the metal is thrown too hard against the far wall of the flask chamber it will recoil, creating a turbulence that can cause irregular hardening within the casting and a loss of detail on its far wall.

Centrifugal force can be supplied in several ways. Probably the most common centrifugal casting machine is driven by a powerful coiled spring. This looks like an overgrown watch mainspring and is kept curled up, out of sight, in the base of the machine. This spring can be wound to great pressures and is potentially dangerous. Never try to remove it from the enclosing base.

↑ TAB GOES HERE

Steps in Centrifugal Casting: Summary

1 Before burnout, balance the casting machine. Check to see if a cradle will be needed to raise the flask into alignment with the crucible.
2 When the flask is burned out, begin cool-down.
3 Wind the casting machine three full turns and brace it with the holding pin. Preheat the crucible, then add metal and bring it nearly to its melting point.
4 Set the flask into the machine.
5 Complete the melt, adding a pinch of flux once or twice.
6 When the metal is molten, grip the arm of the machine and release the pin, keeping the torch on the metal. Check the fluidity of the melt.
7 Release the arm while simultaneously lifting the torch.
8 When the machine has stopped spinning, remove the flask.
9 When the button shows no redness, plunge the flask into a bucket of water.

Detail of the Steps

1. Before burnout, set the flask into the casting machine for a "dry run" to guarantee that the flask will fit in the machine, and determine whether or not a cradle will be needed to lift the flask into alignment with the mouth of the crucible. A cradle is a strip of metal that supports and elevates the flask to ensure a straight line of travel for the molten metal. It is more efficient and safer to figure out these details before the mold is hot.

This is also a good time check the crucible to be sure it is lined with borax, free of contamination, and secured into the casting machine. See below for information on preparing a new crucible. In the case of a previously used crucible, be certain that the mouth of the crucible is not blocked.

Set the metal in the crucible and balance the machine by loosening the knob that holds the arm of the machine onto the base. This becomes the fulcrum of a teeter-totter. Add or adjust weights until the arm balances, remembering that burned out flask is slightly lighter than a fresh-ly poured flask because of the weight of the moisture. Tighten the knob.

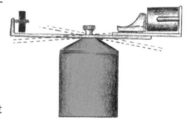

If you deal with similar objects from one casting to the next, it's not necessary to do this each time. Because most rings, for instance, will weigh about the same, once you have established a set-up, simply repeat it with each similar flask.

2. When burnout is complete, turn off the kiln to allow the flask to cool slowly to a temperature about 300-400°F/150–200°C below the melting point of the metal being cast. Large simple objects can be cast in a cooler flask than items with a filigree-like detail.

3. Wind the arm of the centrifugal machine three turns. To achieve this, start with the arm above the holding pin—unwind the arm as necessary until the spring engages with the arm where you want it.

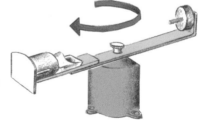

After some experimentation you might find that some castings need more or less force to drive them into the mold. A small object with grillwork, for instance, might need 3 1/2 turns, while a simple large form needs only 2 1/2.

4. Use tongs to remove the flask from the kiln and set it into the casting machine. Remember to turn off the kiln before reaching into it! Avoid jarring the flask during this transfer because its contents are fragile. Especially for delicate models, a bump here could damage details inside the mold.

5. Set the metal into the crucible. A useful trick when using small bits of clean scrap, is to wrap the pieces and a tiny bit of flux in tissue paper. Melt the charge using a reducing (fuel-rich) flame, working as quickly as possible, to minimize the opportunity for oxides to form. Some casters preheat the crucible before adding the metal to shorten the melting time. Do not confuse efficiency with rushing. Use a large but reducing flame, as opposed to the hotter oxidizing flame that will speed melting but might do damage in the form of pits in the casting. If for any reason the arm should slip, simply step back and allow the machine to spin down. Do not try to stop the machine.

6. As the metal melts, it loses its shape and rolls into an oval lump in the center of the crucible. Examine this lump to be sure there are no sharp pieces sticking out from it or coarse rough areas just below the molten surface. Continue to heat until these indicators of solid metal have disappeared.

Check the melt by feeling it with a stirring rod or a stick. Commercially available stirring rods are sticks of carbon or quartz about a foot long and 3/8" in diameter. A length of stainless steel or nichrome wire can also be used, as can a piece of wood. Impurities such as bits of investment or pumice usually show up as glowing orange flecks that float on the melt. Slide them out of the crucible with the stirring rod.

When the metal is completely molten, get a firm grip on the casting arm and pull it in the direction of winding to release the holding pin, which drops when the tension is removed. Be certain to grab a part of the arm that gives a stable grasp; some casting machines have weights on threaded rods that can rotate (unscrew) and slide out of your grip.

Jiggle the arm to test the fluidity of the melt. The metal should roll around in the crucible like mercury. If you see any bulges or solid areas, continue to heat the metal until they disappear.

7. When you are convinced that the metal is molten, simultaneously lift up on the torch and release the arm of the casting machine. This should not be a dramatic gesture: don't jump backward as you do this. It's not necessary and it's not safe.

Allow the casting machine to spin unimpeded. Turn off the torch (oxygen first) and let the machine spin down until it comes to a stop by itself. The metal will probably have hardened by the time the spin is about half spent.

8. When the machine has stopped, remove the flask with tongs and look at the button. The flask can be quenched in a bucket of water as soon as its redness has disappeared.

BUTTON READING

Observation of the button can give important clues about a casting. It might not be possible to rectify problems with the casting just made, but you may be able to figure out how to correct them in future castings.

Observation: Button fills about a third of the sprue former.

> **Meaning:** The correct amount of metal was used.

O: No button; can see into the sprues.

> **Meaning:** Either not enough metal was used, or it didn't get into the flask. If the crucible was clogged or the metal wasn't molten when the casting was thrown, the charge will still be sitting in the crucible. In this case, return the mold to the casting machine and start over.
>
> Another explanation is that you didn't use enough metal. In this case set the half-filled flask back in the casting machine, and repeat the steps outlined above. If successful, this will cast the object in two parts that can be soldered together after they are cleaned up. This is not a sure-fire prospect, but I've seen it work and you have nothing to loose.
>
> Another explanation for empty sprues is that the top of the mold broke through and the metal was thrown out the top of the flask. In this case you'll find a lump of metal on the backplate of the casting machine.

O: The button has lumps and sharp edges sticking out.

> **Meaning:** Either the metal was not completely molten when the casting was thrown, or there are metals with a higher melting point in the button that were floating in the melt and were the last to leave the crucible.

O: The button bridges from the flask to the crucible. (It was probably necessary to break the flask free of the machine).

> **Meaning:** You used too much metal, or the metal did not fill the mold cavity. This could have happened because improper spruing forced the metal to make abrupt turns, or because a piece of foreign matter blocked a sprue.

O: The button has a rounded edge.

>**Meaning:** The metal was starting to harden by the time the button was formed. This might not affect the casting, but indicates that you are being conservative in judging when the metal is molten. Heat it a little longer next time.

O: The button has a thick layer of green glass.

>**Meaning:** You're using more flux than you need to.

The Quench Bucket

Use a plastic bucket; metal ones will quickly corrode through. A hot flask can burn through the bottom of a plastic bucket so for the first few uses, hold the flask suspended rather than dropping it into the bucket. After a couple quenches a layer of white muck will coat the bottom of the bucket and protect it from melting.

The bucket will eventually fill with used plaster that looks like white mud. It cannot be reclaimed and must be dumped into a trash container. Don't get anything too heavy to carry—a standard scrub bucket will weigh about 40-50 pounds when full of investment muck. Disposal is easiest after the investment residue has partially dried, which can take a week or so. Having two buckets in rotation makes a workable arrangement.

Quenching causes splattering, especially as the bucket fills and you're working near the top. Use newspapers or a mat of some kind to protect the area around the bucket.

Because of the low pH of the gypsum-laced water, the quench bucket is very corrosive. A flask left in the bucket for a week can be corroded into a flimsy lace. Scrape off all traces of investment from flasks as soon as soon as possible, working over the bucket or a trash can. Set the clean flask on a shelf to dry.

Machine Guards

 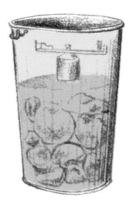 

Set the casting machine inside a washtub and bolt both to a sturdy table.

Secure the machine in a trash can filled with rocks and cement.

Make a table with a space to hold the machine and protect agains splash.

Any centrifugal machine risks the possibility of hot metal being thrown around the studio. This is not an inevitability, but it can and does happen. It is foolish not to plan ahead to protect yourself against injury.

The guard is a wall of sheet metal that surrounds the casting machine and is at least 3" higher than the crucible. These sketches illustrate several possibilities. Because it's difficult to retrieve bits of metal from tight corners,

soften all the floor-to-wall areas with caulk of a strip of duct tape.

If you're a real pioneer, you can build a casting machine from an old bicycle wheel. Either rig up a spring or simply give the wheel a hard spin when the metal is melted. Unless the flasks are very small, add counter-weights on the wheel opposite the crucible. Remember to include a guard.

Sling Casting

In this centrifugal system, the operator supplies the power.

Make a sling as shown, keeping in mind that it doesn't have to be anything fancy. The sizes indicated on the sketch are appropriate for flasks up to about 2.5" x 2.5". This technique is not recommended for flasks much larger than that.

Basic sling casting device.

Spruing must be handled differently because the sprue former, the funnel-shaped opening in the top of the flask, will serve as the crucible. In order to keep the molten metal from dribbling into the mold cavity as it melts, the sprues must be small enough that the surface tension of the metal prevents it from entering the sprue openings through the force of gravity. This is achieved by using round sprues that are no larger than 14 gauge B&S. (1.6 mm) Sprues of rectangular section, cut from sheet wax, can also be used.

Note that because these sprues are small, more than one will probably be needed. It might also be necessary to include on-sprue reservoirs (see Chapter 2 for more information).

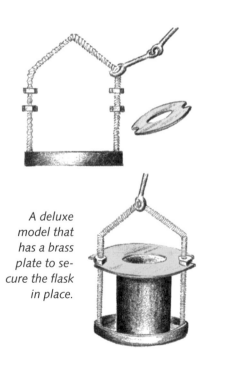

A deluxe model that has a brass plate to secure the flask in place.

Burnout is achieved as in a conventional investment casting. When the flask is ready, set the sling caster onto a heatproof surface such as fire brick and check the area all around and overhead to be sure the swinging action will not be impeded.

Set the flask into the center of the sling's pad. If using the model #2 shown here, clamp the hold-down lid into place by screwing the bolts down the threaded rods.

The Sling Casting Process

Set the metal into the neck of the flask and melt it with a torch, adding flux a couple times. When the metal is completely molten, stand close to the sling, raise it high enough to straighten out the cable or rods, and slowly sway it back and forth a few times to get the feel of it. Keep the torch on the metal while this is going on.

Just before throwing the casting, check to be sure that the line of the swing is free of obstructions (and by-standers). Remove the torch and at the same time swing the sling overhead in a broad circle. Speed is not as import-ant as a constant pressure, in this case not only to force the metal into the mold, but also to keep the hot flask from dropping on your head. After about a minute of swinging, gradually slow down the swing and allow the sling to come back to vertical position. When the button has lost its red color, the flask can be quenched. The process has a certain drama, but when done with a common sense and a cool head it is perfectly safe and practical.

Another Option

A simple centrifugal machine can be made of a couple pieces of pipe that telescope together. Weld the cross bar to the outer tube and configure some way to attach a coun-terweight. Wrap a piece of rope around the column and pull it when the metal is molten. This will cause the pipe to spin and force the metal into the mold as the flask is pulled into a horizontal position. It probably goes without saying that the machine must be securely fas-tened to a solid base and should be surround-ed by a metal shield.

Steam Casting

This simple and effective method makes ingenious use of the existing condition of the burned out mold—its heat—to create a force that will drive molten metal into the mold cavity. It can be used on any size and many shapes of flask, but has limitations on very large pieces because a complicated network of sprues would be needed.

As in sling casting, the funnel-shaped opening at the mouth of the flask will serve as crucible. To prevent melted metal from dripping into the mold cavity, the sprues must be no larger than 14 gauge B&S (1.6 mm). Depending on the sprue former that was used when investing, it might be necessary to create or enlarge the funnel shape. This is easily done before burnout by carving the investment with a knife. If it is to be done after the mold has been evacuated, bits of plaster might fall into the cavity and block the passage.

A steam casting handle is easily made from a large jar lid and a dowel or file handle. Lay a 1/4" pad of newspaper or paper towel into the lid. It's not necessary to be too fussy about this, but make sure there is an adequate cushion of roughly the same thickness all over the cap. Dip the handle in water to completely saturate the paper.

When the burnout and cool-down have been completed, set the flask onto a level and stable heatproof surface and set the metal to be cast into the neck of the flask and melt it with a torch, adding flux a couple times in the process of the melt. Do not use too much flux; a sprinkle of powder will do.

When the metal is molten, pull the torch away and press the casting handle firmly onto the flask. The heat of the molten metal will instantly convert the water in the paper pad into steam. Because steam takes up more space than liquid water, there is a considerable outward pressure that pushes the molten metal into the flask.

Push down on the handle with sufficient pressure to guarantee that the force of the steam being created is directed downward, and hold it in place for about half a minute. When the redness has left the button, the flask can be quenched in water.

raw potato

To make this simple process even simpler, substitute half a potato for the steam handle. Wear a heavy glove to protect against burns. This is not recommended for melts above an ounce.

Vacuum Assist Casting *(also called Vacuum Casting or Pressure Casting)*

This popular casting method uses vacuum pressure pulling through the porous investment mold to draw molten metal into the mold cavity. Among its advantages are the facts that unwanted gases are drawn out of the mold before the metal enters and it's safer for the operator, because nothing is being swung around.

EQUIPMENT

A vacuum pump capable of creating a strong and uniform pressure is needed. The same motor and pump described for investing in Chapter 3 will usually suffice for vacuum casting. Most commercial arrangements have a separate table for investing and casting, but the two functions can take place on one table as long as the proper pads are used.

Investing requires a pad of relatively soft rubber, larger than the base of the bell jar by at least several inches. For casting, the machine needs a smaller pad of heatproof silicon rubber. In a pinch, an 1/8" pad of dampened paper towels can also be used.

Summary of the Process

1 Before burnout, check the seal in the vacuum chamber.
2 Melt the metal in a pouring crucible or a melting furnace adding flux as needed.
3 When the metal is approaching pouring temperature, set the flask on the rubber pad (gate facing up) directly over the vacuum hole.
4 Turn on the vacuum pump and direct the vacuum to the casting table. Check the pressure gauge to be sure the flask has at least 25 inches of pressure. If a seal is not intact, press down on the flask with tongs.
5 Pour the metal into the flask in a smooth even motion.
6 After about half a minute, turn off the vacuum and remove the flask to cool down. Quench the flask when the button has lost its red color.

STEPS IN VACUUM CASTING

1. Before burnout, check the flask to be sure of a good seal and broadly distributed vacuum. To distribute the vacuum throughout the flask, leave the rim of flask exposed, either by remembering to stop before filling the flask all the way to the top or by carving investment away with a knife after it has hardened.

If there is a question of space between the top of the model and the top of the flask, it's possible to carve a recess shape as shown. As a rule of thumb, the mold should be 1/2" thick at the top.

To be sure that the mold will seal, rub the top edge of the flask on a sheet of coarse sandpaper. If there are low spots they will show up as dark unsanded areas in an otherwise shiny rim. Continue sanding until all low spots have been removed. In running water, rinse away any steel filings that cling to the investment. Of course all this is done before burnout.

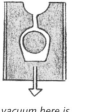
RECESS

The seal can also be checked by setting the mold onto the silicone rubber pad on the vacuum table and turning on the motor. The gauge should read at least 25 inches.

vacuum here is limited   *recess distributes the vacuum*

2. If a self-contained melting furnace is being used, it could have been started while these final preparations were being made. The amount and kind of metal being cast and the size and quality of the furnace will influence how long it takes to melt the charge—it could be from 1 to 10 minutes. If a pouring crucible is being used, set it onto a heat proof surface within arm's reach of the casting machine and begin to melt the metal, adding flux as usual. As the metal melts, add another sprinkle of flux and swirl the crucible to be certain it is entirely fluid.

3. Set the burned out flask on a silicone rubber pad on the casting table. The center of the flask should be directly over the hole through which the vacuum will be drawn. To prolong the life of the pad, protect it with a few layers of dampened paper towel with a hole in the middle.

4. Turn on the machine and watch the gauge to be sure that a seal is made. If the needle doesn't move, press down on the flask with tongs. If this doesn't work, try repositioning the pad, or using its other side. When the melt is complete, lift the crucible into position over the flask and tilt it so the metal is poised on the edge. In a single motion, pour the metal into the mouth of the flask. This should

STEPS IN VACUUM CASTING
(continued)

not be a throwing or jerky motion, but it must not be too slow either. In the case of a torch melt, hold the flame over the lip of the crucible so the metal passes through the flame as it leaves the crucible.

The machine can be turned off within a half minute of the completed pour and the flask can be set aside to cool. Do not leave the flask on the pad to cool because it will cause unnecessary wear to the rubber. When the redness has left the button, the flask can be quenched.

By using the sprue system described earlier for sling and steam casting, metal can be melted directly in the funnel shape of the mold. When the metal is molten, turn on the vacuum.

Vacuum Channels

If the model is thin and radiates out from the center of the flask, it's a good idea to direct the vacuum toward the outer edges of the mold so it can pull the metal outward. To do this, drill holes (about 1/8") into the hardened investment as shown. Be sure to clean and dry the drill bit or it will soon rust. An alternate method is to drape steel rods (coat hanger works well) over the edge of the flask before investing.

These are pulled out before burnout. The rods can also be made of wax or soda straws, in which case they will burn away. To use the straws, cut them 1/2" shorter than the flask height and plug the bottom end with wax. Immediately after filling the mold with investment, slide these into place and secure them with bobby pins. Take the pins off before burnout.

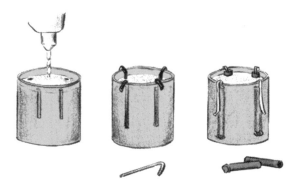

Perforated Flasks

For large castings, a special flask is used to completely surround the mold with vacuum pressure. This vessel is pierced with many holes, and in fact is not so much a flask as an armature to support the investment. During moldmaking, a plastic or rubber sleeve contains the investment and is stripped off after it has hardened. As shown, the flask is set into the top of a specially built vacuum chamber where a powerful suction pulls at the mold from all sides. A vacuum chamber like this can be improvised from a flask (or a coffee can) an inch or two larger than the perforated flask being used. Use a sheet of steel or aluminum across the top to seal the makeshift vacuum chamber.

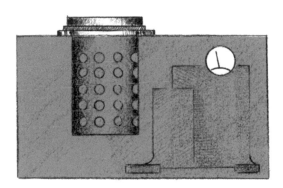

CLEAN UP

When the casting is completed, plunge the still-hot flask into a bucket of water where it bubbles and hisses as the investment breaks away from the flask and the casting. The steam being released in this process is laden with silica particles and therefore very nasty to breathe in. Use ventilation and a respirator when quenching.

Clean the flask by scraping it with a knife (an old butter knife works well), and set it aside to dry. With your fingers or the same knife, remove loose investment from the casting.

In a commercial shop, castings are cleaned either by holding them under a jet of high pressure steam, or by using the microscopic scrubbing action of an ultrasonic machine in a solution sold for this purpose.

In a small shop, castings can be cleaned by scrubbing them with a toothbrush. This is done first in the investment bucket, where most of the residue is brushed away, and then in the sink for a final cleaning. When all traces of investment have been removed from the casting and button, both are cleaned in a warm solution of pickle.

Not all castings require models, and in some cases the mold is so vague that the term hardly applies. The casting methods described in this chapter are usually primitive, both in historical and aesthetic terms. They are simpler than investment casting and involve less equipment and fewer steps.

The Pouring Crucible

This is a simple ladle made of a ceramic bowl and a steel handle, available through most of the casting suppliers listed at the end of this book. A separate crucible should be maintained for each metal being cast.

Crucibles can last a long time but will eventually crack, so I recommend having one in reserve. Broken edges pose no threat, but you should discard the crucible at the first sign of a center crack. Better safe than sorry.

Between uses, keep crucibles tightly sealed in a plastic bag so they don't absorb moisture. Dampness in the cru-

cible will turn to steam when heated, and this can crack the crucible. Before the first use, line the crucible with a coating of borax glass to seal the pores of the crucible and block the entrance of oxygen from beneath the metal.

To do this, set the crucible on a fireproof surface and heat it with a torch until the bowl is a glowing red. Shift the torch off to one side and sprinkle borax powder into the bowl, repeating this several times until the interior of the crucible develops a glassy layer. Allow the crucible to air cool, away from drafts.

TAB GOES HERE →

Homemade Ladle

Casters with a pioneering spirit might want to make their own crucibles from investment. See Chapter 3 for a description if you are unfamiliar with this plaster-like material.

Mold a lump of clay into a mound that will be the shape the bowl of the ladle. Set the clay onto a smooth surface and build a wall around it with clay, plastic or cardboard, then mix investment as described previously. To give resiliency to the crucible, stir in powdered pumice, adding about 1/3 to 1/2 as much pumice as there is investment. Pour this mixture into the form and tap it to remove air bubbles. Allow the crucible to dry overnight, then pull off the

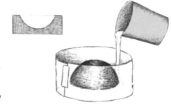

form and bake it at 600oF. Allow it to air cool slowly and it's ready for use.

Firebrick Ladle

A simple alternative to a conventional crucible is a hollowed out firebrick, fastened into a handle or held in tongs. In some molds, it's possible to carve a crucible adjacent to the mold cavity, in which case "pouring" consists of simply tilting the mold when the metal is molten.

Handles can be made of any rigid metal strap. Aluminum molding about an inch wide and 1/8" thick is available through hardware stores and has the advantage of being easily shaped and drilled to accept a bolt. A round rod (e.g. welding rod) can also be used but I recommend forging the handle area flat so you'll have a positive grip.

Casting into Water

Some interesting effects can be achieved by pouring molten metal into water. The results will be altered by:

- the temperature of the metal.
- the depth and temperature of the water.
- the kind of metal being used.
- the quantity being poured.
- the speed and direction of the pour.

Heat the metal in a pouring crucible or a hollowed-out firebrick, adding flux once or twice. The water must be deep enough to allow the metal to cool completely before it hits bottom. If the bucket is too shallow the metal will stick to a metal bucket or melt its way into a plastic one.

Stand with the crucible at least two feet above the bucket and let the molten charge dribble into the water, experimenting with large and small dribbles, and with various movements of the crucible as the pour is made. Alter the distance between the water and the metal to discover a range of effects. Sort through the results, select out the best shapes, and remelt the rest.

Experiments can be made by pouring into ice water, brine, water that is moving, boiling water, and so oe. Pouring into cold still water will produce the "corn flakes" characteristics of this method. These offer potential as components in fabricated jewelry pieces but they are not recommended as casting shot because their large surface area risks serious oxidation during remelting.

A better method for making shot is to drill several small holes in the bottom of a new crucible, using a masonry bit. Heat the metal until it drips through, adding more metal to the charge as you proceed. Let the metal drop at least two feet into hot swirling water. This will result in small spheres that are easily remelted. Never pour into any liquid that is flammable.

OTHER POUR MEDIA

- sawdust
- rock salt
- sand
- rough wood (hollows and knots)
- soft dirt
- candy

The important factor in all of these materials is that any pieces that get trapped in the metal can be removed, usually through dissolving or burning. It's possible, for instance, to pour onto gravel, but when the metal surrounds a piece of the gravel it will entrap it, making it permanently connected to the piece.

CARVED STONE

For many years, Native American jewelers made gravity castings into a naturally occuring stone called tuff or tufa that consists of compacted volcanic ash. The stone is soft enough to be cut with a carpenter's saw and can be worked with nails, old files, or bits of wood.

As a substitute, lightweight firebricks can be used. Some of these a have large sponge-like holes that will show up on the casting as lumps, making a clean design almost impossible. For this reason, use only a brick with small spaces.

Charcoal blocks are a better alternative. The biggest disadvantage here is the cost, which is high, especially when making large castings. For small parts, however, where several pieces can be fit into a single block, the cost is reasonable. Charcoal is harder to carve than tufa, but it will reproduce texture nicely.

Alternate Casting Blocks

Another substitute can be made in the shop with investment or molding (regular) Plaster of Paris mixed with powdered pumice.

On a flat surface (a sheet of glass or Plexiglas is ideal) lay down sticks to enclose an area and tack them into position with clay. Mix investment or plaster as described in Chapter 3 and remove air bubbles by tapping, vibrating, or in a vacuum. Pour the mix into the form and allow it to set up for at least an hour. The sticks can then be pulled away, and if desired, the mold can be scored for breaking. Use a knife and straight edge to scratch a line into the plaster. Then slide the block off the flat surface and break it over an edge.

Curing

Allow the mold pieces to dry thoroughly. They can be air dried or the process can be hastened by using sunlight, a lamp, or a kiln set on low. But they MUST be dry. Even a small amount of moisture will be instantly turned to steam as the molten metal makes contact with the mold and this steam will press outward against the molten metal, preventing it from assuming the shape of the mold. In some cases it may also cause the hot metal to "spit" out of the cavity, which can be dangerous.

Depending on the complexity and scale of the casting, and assuming there are no undercuts, a mold like this can be used for as many as a dozen castings. Eventually the detail will break down, so each casting is a little more distorted or requires more finishing than its predecessor.

Process for a Pressed Casting (aka Horizontal Casting)

1   Create a flat surface on the mold material (plaster, charcoal, tufa, etc.) by rubbing it on coarse sandpaper.

2   Carve, scratch or press a hollow into the mold. The depth of the mold will become the thickness of the final casting.

3   In the case of a dense material and/ or large cavity, cut vents radiating outward from the design.

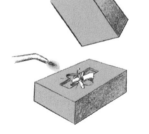

4   Set pieces of metal into the cavity and melt them with a torch, adding flux sparingly midway through the melt.

5   When the metal is freely molten, remove the torch and promptly press down on the lump of metal with a flat fireproof material, such as another mold block. Hold firmly for about 30 seconds.

NOTE: THIS IS A POTENTIALLY DANGEROUS OPERATION. STAND RATHER THAN SIT SO YOU CAN MOVE AWAY IF MOLTEN METAL SQUIRTS OUT OF THE MOLD. WEAR A FIREPROOF APRON AND GLOVES.

Process for a Poured Casting (aka Vertical Casting)

1   Prepare two mold pieces with perfectly flat sides by rubbing each piece on sandpaper.

2   Carve the impression of the casting into one block. Casting an object with pattern on both sides is possible but registration (lining up the two parts) is difficult. You can solve this by making one side a random pattern.

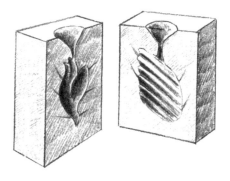

3   Scratch radial vent lines.

4   Carve a sprue channel from the heaviest part of the casting to the outside edge of the mold and enlarge this to a funnel at the top. This sprue should be no longer than an inch.

5   Tie the two mold pieces together with wire.

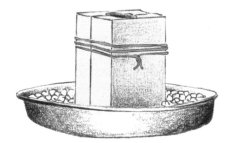

6   Prop the mold in a pan and warm it slightly with a torch flame. Melt the metal in a pouring crucible, adding flux once or twice. When it is completely fluid, pour the metal into the mold with a smooth movement.

Cuttlefish Casting

I shudder to think what led the first metalsmith to try making a mold from the skeleton of a squid-like mollusk. Whatever it was, the results were interesting enough to warrant further experimentation and the result is a versatile and interesting casting method.

MATERIAL

Cuttlefish skeletons are composed of a chalky white material soft enough to be scraped with a fingernail covered on one side with a dense plastic-like layer. The skeleton shows the growth pattern of the animal in a series of irregular lines that resemble woodgrain. Small cuttlefish bones are available through pet stores, where they are sold for parakeets, who use them to sharpen their beaks. The larger bones preferred for casting are sold through jewelry supply houses, usually for about a couple dollars a piece.

The following description assumes a large bone, used for both sides of the mold, but many variations are possible, including the use of other materials for the back of the mold.

Cuttlefish Process

1 Cut the tips from a large bone with a jewelers or a coping saw.

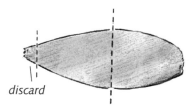

discard

2 Flatten the soft side of the bone by rubbing it on sandpaper or a rough surface such as concrete.

3 Cut the bone in half. Continue flattening by rubbing the two pieces against one another in a circular movement. This is messy, so work over a trash can.

Rub pieces against each other to flatten them.

4 Carve or press the image into the bone, blowing away the dust as it forms. If you want to accentuate the grain pattern, gently brush the cavity with a dry paintbrush.

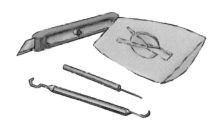

Use a soft paintbrush to exaggerate the natural texture

Cuttlefish Process *(continued)*

5 Carve a sprue or gate from the thick part of the casting to the top of the bone. Enlarge this area to a funnel shape. The sprue should be no longer than an inch.

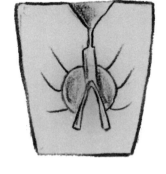

6 Scratch thin radial vents into the mold to allow air to escape.

7 Tie the two parts of the mold together with wire and prop the mold up in a dish of sand. This is not essential to the process, but it is handy and will catch any spilled metal safely.

8 Melt the metal in a crucible and pour it into the mold. There will be a characteristic odor, the smell of, well, burning fish skeleton. Allow the metal to cool for about a minute, then open the mold and remove the casting with tweezers. Cuttlefish molds can be used only once.

Texture

The rich linear texture of cuttlefish is what makes it so appealing and at the same time so difficult to use well. While the rich texture is a tribute to Mother Nature, it is not automatically good jewelry. To make use of the texture as a designer, think in terms of contrast between heavily textured and smooth areas that can be created by filing away or planishing selected areas of the finished casting, or by using cast cuttlefish sections in conjunction with fabricated sheet.

USING A MODEL WITH CUTTLEFISH

This technique can also be used to duplicate an object that cannot be burned out for investment casting. A model can be carved in wood, for instance, or an existing metal casting can be duplicated. For this process, the model cannot have undercuts.

1   Begin as above, being sure to use a piece of cuttlefish that will allow for at least 1/4" thickness of bone all around the model. Prepare the sides, making them so flat that when held together no light can be seen through them.

2   Determine the orientation of the model and the place where the sprue will connect with it. Press the model halfway into the bone.

3   Press the other half of the mold onto the model, distributing the pressure as broadly as possible to avoid cracking the fragile bone. Press until the two pieces are touching all along their flat surfaces. Use a file or a felt marker along the outside of the two pieces to assist alignment of the mold halves.

4   Gently pry the mold open and lift out the model. If the grain is to be accentuated, brush the mold cavity with a soft brush.

5   Use a knife blade or similar tool to carve a sprue and gate into each side of the mold, being careful that no pieces of bone lodge in the mold cavity. Scratch some radial vent lines from the heaviest parts of the casting.

6   Lay the halves of the mold together, being careful to line up the notches. Tie the assembly with wire and prop it in a pan.

7   Melt the casting metal in a pouring crucible, adding flux twice. Pour the molten metal in a relaxed and uninterrupted flow.

Making a Three-Piece Cuttlefish Mold

Complicated models can be accommodated with a three-part mold. The lower sections of the ring shown here are treated as described previously, positioned so the top section projects above the mold halves. After cutting notches, the model is removed and the top of the mold is filed smooth. A thick piece of cuttlefish is filed to make a close fit against this surface.

The mold is then opened and the model is carefully set back into place. The top mold piece is pressed onto the model and registration marks are made, either in the form of notches or lines with a marker. The mold is then dismantled and the model is removed. A sprue is cut and the pieces are tied together with wire. Though the development of rubber molds have largely replaced this method of duplicating metal objects it was once a common practice among jewelers.

Registration notches.

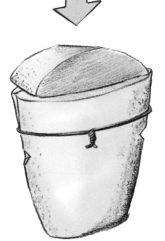

CHAPTER 7

SAND CASTING

Advantages

- low cost
- simple equipment
- versatility

Design Considerations

Sand casting cannot be used easily with shapes that contain an undercut. Because the model must be mechanically removed from the sand mold, lifting out an undercut shape will tear away a part of the mold. Complex multi-part molds can sometimes be devised, but in this age of rubber molds and investment casting, complex shapes are usually cast with those techniques.

    This is not to say the sand casting is outmoded—far from it. Literally millions of objects are sand cast each year, from small engine parts to huge architectural components.

Materials and Equipment

- fine sand, also galled green or French sand
- mold frames, cope and drag
- heat source for melting the metal
- a couple flat sheets of glass, plastic, or wood (called molding boards)
- pounce, a dry mold release
- a sieve, wooden dowel or block for ramming, and tweezers

←—— TAB GOES HERE

Models

Models must be able to stand up to the pressure of hard tamping and are typically made of metal, plastics or close-grained woods such as mahogany or cherry. Any materials can be combined; glue, screw, or solder the pieces together.

Wood models should be sanded smooth and painted with several coats of shellac or varnish to seal the surface. Avoid perfectly straight sides, carving models with a slight bevel, called a taper or draft. Models can be used for hundreds of castings, so it makes sense to devote whatever time is needed to create a perfect model.

Sand

As might be expected, this is the most critical ingredient in sand casting. The size and shape of the particles of sand will affect the detail of the casting, and in fact the actual success or failure of the process.

The finer the sand, the better the detail on the finished casting will be. In addition, fine sand is needed to hold together tightly when the mold is being made. It is possible, or course, to find a local deposit of sand and sift out the finest particles, but a more realistic solution is a contact with a commercial supplier. This is not as easy as it once was. Since investment molds have come to dominate the jewelry industry, many suppliers have stopped carrying top quality sand. One possible source is through local foundries, which probably buy their sand in large quantities and could spare the small amount a jeweler will need. In a pinch, fine pumice can be used, but proper casting sand is preferred. Beach sand does not make good molds because the action of the sea has worn the particles to round shapes that do not hold together well.

NATURAL SAND

This describes a large category of casting sands that are used pretty much in the condition in which they are found in nature. They are a mixture of silica sand with some amount of clay, usually between 10 and 30%. These sands are named for their source; a couple of popular types being Albany Sand and Greely (Colorado) Sand. French Sand (yes, from France) contains 16.5% clay and is available in mesh sizes from 135 to 170.

Benvenuto Cellini, writing in the 1560s said, *"The clay from France at Paris is the finest I ever saw. As a rule the clay from the grottoes is better than that from the rivers."*

SYNTHETIC SAND

This is a misleading term because all the ingredients are natural. The synthesizing is a matter of combining ingredients that don't usually occur together. The sand used is a sharp, clay-free silica sand. The clay is a fine alumina silicate called bentonite, named for its source in Fort Benton, Montana.

Many commercial foundries use a mixture of sand and clay (and sometimes resins) called foundry sand. Molds of this material are made as described above, but are then fired in kilns like ceramic ware. This yields molds of great strength, but is a process better adapted to large scale commercial castings in steel than to jewelry needs. If a jeweler were going to get that involved, it would be easier to pursue investment casting.

A reasonable shortcut is a process called skin-drying. This toughens a mold and if done correctly yields an exceptionally smooth surface. After making a sand mold in the usual way, the interior of the mold is sprayed with a mixture of 10 parts water and 1 part molasses using a fine spray mister. After air drying, warm the mold with a gentle torch flame to set the sugary mix. Casting is done as usual.

Dampness

In order to bond the sand into a mold, it must be dampened (*tempered*) with either water or oil. The advantage of water is that the mix can easily be dried if too much liquid has been added. The advantage of oil is that it will not dry up between uses.

To prepare a water-bonded sand, spread the sand over a board or in a low flat box and spray clean water over it with a plant mister or spray bottle. Proper moldmaking uses as little water as possible, something under 5%. Knead the water into the sand by repeated mixing with a paddle or by hand. Continue to add water sparingly until a handful of sand, when pressed in a fist, retains every detail of the handprint. The mix does not feel wet: if sand particles stick to your hand, the mix is too wet. If the edges of the lump crumble, the mix is too dry. A pressed lump of the correct consistency will be firm enough to stay together even when dropped back into the sandbox.

Even though this looks good, the sand is probably slightly wetter than need be. If possible, mix the sand the day before it is to be used to allow the water to spread through the sand to achieve the proper consistency. This will allow the water to find a more complete distribution and dry out slightly at the same time. If too much water was added, spread the sand out on a board and leave it in the sun to dry or use a hair dryer if you're in a hurry.

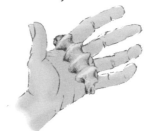

OIL

A common oil for bonding casting sand is glycerin, which can be bought at most drugstores. It is worked into the sand as described for water.

Water-based sand will dry out in a day or two, and will need to be redampened before the next casting. An advantage of dampening sand with oil is that the sand is dried only through use, when the heat of the molten metal burns the oil away. It will have to be rejuvenated eventually, but much less frequently than the water-based sand.

Because it cannot be dried, mixing too much oil into the sand can present a dilemma. To allow for this, keep some dry sand in reserve. If the mix accidentally gets too wet, the dry sand can be sifted in to correct the balance.

Rather than clog up a spray bottle, it's possible to sprinkle the oil over the sand, but a little more mixing will be needed to distribute the liquid throughout the sand. Sift and resift handfuls of sand to hasten this. A couple of young children can assist very well. By the time the fifth castle has been made and destroyed, the sand will be pretty well mixed.

It is not necessary to wait with oil-bonded sand, but it is beneficial if the mix can sit overnight. Slightly warming the sand will help disperse the oil.

Cope and Drag

Mold frames are usually made of cast iron, and are available in several sizes and shapes from suppliers of casting supplies. The halves are identical except that the cope has pins that slide into lugs in the drag to align the halves of the mold. The complete mold frame will cost at least $40, depending on its size. This is a good investment, because the frame will last indefinitely. Extension rings called cheeks are used to make tall molds, but these have very little use for jewelry scale casting. Mold frames can be made of hardwood

as long as the gate is metal. Wooden frames are not as good as metal ones because of the possibility of warping and the fact that spilled metal will damage them, but for the enterprising caster looking for a low cost introduction, homemade frames are a workable alternative.

It's important that the frame has a lip to hold the sand in the mold. See the detail for a suggested cross section. The gate in this case is made of short section of steel pipe, bolted onto the frame.

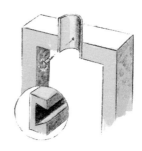

POUNCE

Several powders can be used to prevent the sand between the mold halves from sticking together. Talc, cornstarch, powdered mica, or any dry commercial mold release product will work well. Guard against using too much pounce because it will contaminate the sand and shorten its use.

To make a convenient applicator, cut 3 or 4 pieces of cotton fabric (e.g. an old sheet) about 8 inches square. Layer these and pour a mound of powder in the center, then gather up the corners and tie them together to make a bulb

will last for years before it needs to be refilled. To sprinkle out a fine dusting of powder, tap the fabric against a finger held just above the sand.

Process for a Flat-Backed Casting

1   Prepare the sand to a proper consistency, one that just holds its shape
when squeezed.

2   Spread newspaper over the work
table. This will help in picking up
excess sand and pouring it back into its
container. Set the drag upside-down
on a clean flat surface such as a piece
of glass or Plexiglas. The edge that
joins the other half of the mold will be
facing down.

3   Sift (*riddle*) sand into the mold until
it is about half full and tamp the sand
down with a wooden ram. Start with
light strokes, but gradually increase the
force of your blows until the sand is
very tightly packed.

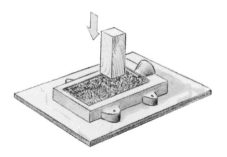

4   Sprinkle in more sand and tamp
again. Before adding more sand, rough
up the packed surface to provide a
key for adhesion between the layers
of sand. Continue building the mold,
layer by layer, until the sand is slightly
higher than the top of the drag.

5   With a piece of metal or hardwood,
strike off, or level the top of the sand.

6   Set a piece of plastic or wood
(called a moldboard) on top of the
frame and lift the drag up, sandwiched
between two moldboards. Turn the assembly over so the new moldboard is
on the bottom and remove the original
board so that the flat surface of the
sand faces up. Set the packed frame
down on a clean hard surface.

Process for a Flat-Backed Casting *(continued)*

7 After lining up the two halves, set the cope onto the drag and dust a thin film of pounce (parting compound) over the sand.

8 Lay the model in position, oriented so its thickest section is between one and two inches from the gate. Dust the model with pounce.

9 Sprinkle/sift sand into the top half of the mold, covering the model and tamping it down as before. Work gingerly, especially at first, to avoid shifting the model then add more sand and pack it down, each time pressing a little harder. Continue until the sand is above the top edge of the frame. Strike off as before.

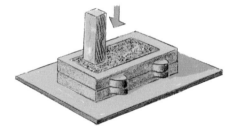

10 Set a moldboard on top of the mold and invert the whole works. Set the packed mold down, again being sure it is level and solidly on the table.

11 Lift off the drag, moving gently. The sand will stay packed in the frame, parting at the powdery layer between the mold halves. Set the drag off to one side.

12 Lightly tap the model to loosen it, then use tweezers to lift it out of the sand. Some casters predrill a hole and use a screw lightly twisted into place to act as a handle. Either way, early training at Pick-Up-Sticks will prove valuable here.

Process for a Flat-Backed Casting *(continued)*

13 Use a knife or wooden tool to carve a sprue from the model to the gate. Carefully brush and blow bits of sand away. A small rubber bulb is handy for puffing away grains of sand that fall into the mold.

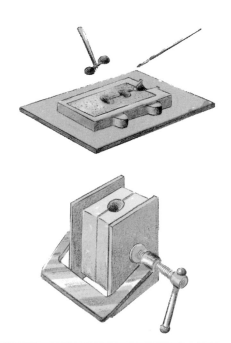

14 Set the mold halves together, still with their wooden supports. Hold them together by lightly clamping them in a C-clamp. Set the completed mold upright in a dish of sand.

15 Melt the metal in a furnace or with a torch, and pour it into the mold. After a few minutes the mold is "shaken out" or broken open. The sand will be darkened and dried out, but otherwise unharmed. It can be used for many years.

TWO-PART MODELS

Complicated pieces, those without a flat or nearly flat back, require a somewhat different approach. This process is widely used in industry but has less appeal for jewelers. It is still the most practical method for some designs, especially when several dozen or more identical pieces are being cast. The design must have no undercuts. The model must be constructed in such a way that it comes apart at the *parting line,* the division between sections. Depending on the material used, the scale of the work, and the equipment at hand, this can be achieved by carefully fabricating two parts independently or by making a model and then slicing it in half. The two halves are held in alignment with pins. The pattern pieces can also be fixed onto each side of a board called a match plate.

Process for an Irregularly Shaped Casting

1   Prepare the sand as described above and get
    the work area ready by laying out sheets of
    newspaper. This will make it easy to collect ex-
    cess sand and return it to its storage container.

2   Set one part of the model (A) onto a sheet
    of wood, plastic, or glass and place the drag
    around it. The parting face of the model
    should be resting on the glass. Dust the
    model lightly with pounce.

3   Sprinkle/sift sand over the model, packing
    it tightly layer by layer. Continue until the
    sand is packed up higher than the top edge
    of the frame, then strike off the mold and set
    a board on top of it. Invert the sandwiched
    mold and set it onto the work table, making
    sure it is level and stable.

4   Slide the other half of the
    model (B) into position, using
    the pins to line up the two
    halves.

5   Set the cope into place and coat the model
    and flat face of the sand mold with pounce.
    Sprinkle/sift sand over the model and tamp
    it down, again starting lightly but using in-
    creasing force until the mold is tightly packed.
    Strike off to achieve a flat surface. If vents are
    needed, use a small wire to poke through the
    sand to the model.

Process for an Irregularly
Shaped Casting *(continued)*

6   Invert the mold and gently lift off the drag, setting
    it to one side. With a knife or similar tool, scrape
    sand away to make a channel (the sprue) from
    the model to the gate. In small castings this can
    be in just one side, but for large masses of metal,
    it might be necessary to carve channels in each of
    the mold halves.

7   Use tweezers or pull-screws to
    lift each half of the model out of
    the sand.

8   Set the two frame pieces back
    together, gripping them lightly
    in a C-clamp and standing them
    upright.

9   Melt the metal to be cast and pour it into the
    mold. After about a minute the mold can be
    opened and the casting may be removed.

Alternate Sprue Forming Methods

When the mold is tightly packed and the model is still enclosed within it, slide a piece of thin-walled tubing through the gate until it hits the model. When this is withdrawn it will bring a core of sand with it, leaving a clean sided tunnel through which metal can be poured. When the model is removed the point of contact between sprue and model might need to be lightly brushed to remove bits of sand that would otherwise fall into the cavity.

It is also possible that the sprue should enter the mold from a side, rather than the top. This depends on the shape of the model more than anything else. This method is not recommended for thin molds, because the weight of the molten metal can cause the sand to bulge out, distorting the object. Where there is no other alternative, use a molding board with a hole.

DIAGNOSING PROBLEMS

Every casting has something to teach us, even the ones that don't come out as well as expected. Of course it's frustrating, but take advantage of the opportunity to investigate what happened and seek a solution.

Observation: the mold falls apart when lifted.

Modification: add more water or oil to the sand. Don't overdo it. A small amount of moisture can make a big difference in the sand.

Observation: The mold halves stick together when being lifted apart to remove the model.

Modification: The sand is probably too damp. Allow it to dry (water) or sift in more sand (oil). In the meantime, you might try using a little more pounce.

Observation: The surface of an otherwise good casting is rough.

Modification: The metal may be too hot when poured, or the sand may be so damp it is creating steam as the metal hits it. Allow the sand to dry slightly.

Observation: The sand always caves in when the model is removed from the mold.

Modification: Check the model to be sure it doesn't have undercuts. See that the sides have a slight draft or slope. Dust the model with pounce. If the problem persists, clean the model well and paint it with shellac or enamel. A smooth, hard, waterproof surface will resist clinging to the sand.

CHAPTER 8

CASTING PEWTER

Combinations of zinc, bismuth, cadmium, tin and lead are known collectively as white metals. Perhaps the best known member of this family is pewter, an alloy consisting mostly of tin with various other ingredients. In the past, an important ingredient was lead, which brings with it health concerns, but modern pewter contains no lead.

Pewter and white metal alloys lend themselves well to casting. They melt easily, stay molten a relatively long time, and fill molds readily because of their low degree of surface tension. While the techniques previously discussed can all be used for white metal casting, this chapter explains variations specially adapted to these low melting alloys.

Common Alloys

% Tin	% Lead	% Antimony	% Other	Melting Temp. °F
92	8			464-471
93.5	3.5	1.5	1.5 Cadmium	458-464
92	4	4		458-464
91		7	2 Copper	560
94	4	2		525
88	6	6		464

MELTING

Pewter and white metals can be melted on a stove or hot plate, in a kiln or with a torch. The best container for melting is a cast iron ladle, available through plumbing suppliers. In a pinch a cast iron skillet, a steel sauce pan or a stainless steel soup ladle will all suffice. A disadvantage of these makeshift solutions is their awkward balance. Pewter is heavy and can easily spill from an improvised crucible. Be especially careful when using these.

It is not necessary to use flux when melting pewter. A dark gray oxide called dross forms as the metal melts, and this film should be skimmed from the surface of the melt before the metal is poured. The dross can be collected and returned to a refiner, but unless large quantities are involved it won't be worth it. When the ladle is empty of clean metal, tap out the cinder-like dross onto a brick or board and throw it away after it has cooled. Airborne dross particles can be toxic so a respirator is advised.

Melting Pots

For large quantity melts, a vat is used to keep several pounds of pewter molten for hours at a time. In this case the dross is allowed to form on the top of the pot and simply pushed aside with the ladle when pewter is being dipped up for a pour. Melting pots can be purchased through some distributors of metalworking equipment, or can be rigged up from a heavy steel vessel and a gas burner. In this case, the pot should be clamped in place as shown to prevent the possibility of spills.

Pouring Crucible

A cast iron pouring crucible has a clever arrangement designed to simplify the skimming operation. The ladle is divided into two sections by a steel or cast iron wall that leaves a space around its lower edge. One side of the ladle is used for melting so the flame is directed only at this compartment. When the metal melts, it flows underneath the divider, where it stays clean and ready to pour.

These are not readily available, so you may want to rig up your own divided ladle. A standard cast iron ladle can be converted by making a divider as shown from sheet steel. Because it is difficult to weld to cast iron, the divider in this design is held in place with tabs.

Remelting

If a large quantity of pewter or white metal has solidified in a pot, special care should be taken when remelting. If pressure that is created as the metal melts is trapped beneath a cap of solid metal, there is a risk of "geysering." Prevent this by heating from the top, for instance with a torch. If a hot plate or gas ring is used, tip the pot so melting takes place on the side.

Contamination

When pewter or any of the low melting alloys are heated beyond their melting point in the presence of other nonferrous metals, a nasty contamination will occur. White metal attacks silver, brass, bronze, copper, or gold with a corrosive zeal. The result will be large irregular pits and eventually holes eaten all the way through a piece of metal. The action is especially insidious because it continues to advance each time the metal is heated. Once a white metal has attached itself to a jewelry alloy, it's almost impossible to clean off and it is usually necessary to cut away the contaminated area. In cases where this is not possible, the whole object must be scrapped. Refiners of precious metals can reclaim the noble metal, but there will, of course, be a charge for this service.

By far the best solution is to prevent contamination from happening in the first place. If you must use the same work area for pewter and other metals, develop a strict routine of clean-up between uses. Contain all the pewterworking equipment in a pan or tray. In this way the pewter, its spills, and any other debris can be removed entirely when not being used. And even so, the tabletop should be swept and the files brushed out before starting to work with precious metals.

The soldering area is especially vulnerable to contamination. The best arrangement provides separate areas for heating white metals and precious metals. If this is impractical, devise a foolproof system that protects against contamination.

PLASTER MOLDS

Pewter and white metals can be cast into molds made of the plaster available through hardware stores and lumber yards. Casting investment can also be used, but it's more expensive and includes ingredients that are not needed when casting pewter.

Mold blocks can be made by pouring a sheet of plaster onto a smooth surface such as glass, Plexiglas or Masonite. Use small boards braced into position with bits of clay as shown to form walls.

Mix the plaster in a flexible container, using about the same amount of water as the desired amount of plaster. The water should be at room temperature or slightly warmer, but not hot. Sprinkle plaster into the water gradually, working until a dry island forms in the bucket. At this stage, reach in with your hand and slowly stir the mix. Avoid rapid and violent stirring—it is better to massage the mix than to whip it into a froth.

With a board or a screwdriver, rap the side of the bucket sharply many times to jiggle air bubbles to the surface. Pour the plaster into the mold frame to a thickness of between a half inch and one inch and spread the surface with the palm of your hand to smooth it.

Allow the plaster to harden undisturbed for at least an hour. Plaster is exothermic and will give off heat as it cures. If the sheet is to be divided into several casting blocks, use a straight edge and a knife or screwdriver blade to score grooves into the plaster. The sheet may be allowed to cure further, or may be broken up at this time.

To break the sheet, slide the plaster off the flat surface so that a scored

line is directly above the table edge. A sharp downward tug will break the sheet along the edge. This is very efficient and a handy trick when you're making many blocks at a time.

The plaster must be **completely** dry before the molds can be used for casting. You can, however, begin the carving process when the molds are damp. The trade off here is that the plaster cuts easier when damp but tends to crumble away, making detailed work difficult. The ideal solution then is to do rough carving within 24 hours after pouring the plaster, especially where a good deal of plaster must be cut away, and then finish the mold several days later. In practice each person's work methods and available time will determine the sequence.

Carving

Plaster molds can be carved with any rigid tool—nails, screwdrivers, drill bits, knives and dental tools will all work well. Remember that the low pH and high moisture content of the plaster will cause tools to rust quickly unless they are washed and dried immediately after use. The easiest way around this problem is to use tools that are expendable.

Designs can be drawn directly on the plaster with a pencil. Sketches can also be transferred with carbon paper or by creating a thick layer of graphite on the back of the sketch and tracing it onto the block. Especially where letters or numbers are being used, remember that they must be carved backwards to come out reading correctly.

The depth of the recess will translate to the thickness of the casting. Pewter is not as strong as sterling, brass or copper so craftspeople familiar with jewelry metals should make a conscious effort to work thick. Of course the most important factor in determining the scale of a piece is the strength it will need when in use.

As the carving proceeds, you can get an idea of how the design is shaping up by making a quick test with clay. Blow or brush away bits of plaster and press a lump of clay into the recess (either earthen clay or plasticine can be used). If you have borrowed clay from a potter, you should be aware that even small traces of plaster can cause big problems in firing, so the potter would probably rather not have his clay back if it has been around plaster.

If the piece has no undercuts, you can make a pewter  casting by way of a trial impression. Because of the low melting point of the metal, the mold can be used many times without deterioration. This means it is possible to carve a recess, then make a casting with the understanding that you can work further on the mold. Keep in mind that further carving, will increase the thickness of the object. If the preliminary checks indicate that you have carved too deeply (i.e. that the cast piece is too thick), sand down the face of the block and in this way diminish the depth of the recess.

Poured (Gravity) Casting

This description refers to flat-backed objects.

To pour a mold, you must first carve a sprue or gate into the recess. The sprue should connect with the thickest part of the casting and should spread outward like a funnel as it nears the outside edge of the block. It can be carved with a knife or any similar rigid tool.

1. Set a flat block against the mold to serve as a back for the piece, and use a C-clamp or wire to secure the two mold halves together. Set the assembled mold into a dish of sand, gravel or a cast iron skillet to support the mold and contain any pewter that spills. By positioning the flat block slightly higher than the mold, a shelf is created to help guide the pour into the sprue.

2. Heat the pewter and pour it smoothly into the mold until it fills the gate. Allow the pewter to cool thoroughly before opening the mold; depending on the size of the piece this can take from 1 to 5 minutes. As the button cools, the surface will change from a shiny to a matte surface. Because this is usually the thickest part of the casting, you can assume that when the button has hardened, the rest of the casting is also solid. Open the mold and remove the casting with tweezers—it is probably too hot to touch.

Poured (Gravity) Casting *(continued)*

If the casting was incomplete or lacking detail it could be because air trapped in the mold cavity prevented the metal from completely occupying the carved recess. Allowance for this can be made in several ways.

(a) Especially in delicate castings, sprinkle a little light powder (talc, corn-starch, borax) onto the mold halves before clamping them together. This will allow a tiny air space between the two parts and provide a vent for the air inside the mold.

(b) Scratch tiny lines outward from the mold as shown. These are generally too small to be filled with metal, but will allow a space for the air to escape.

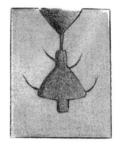

(c) Carve a sprue-like channel called a *riser* as shown here. In this case the metal fills the mold and pushes up through the riser. This is especially useful for large castings and has the added advantage of providing a supply of molten metal to assist in the shrink-ing or recrystallization process that accompanies cooling.

riser ——

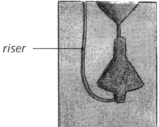

Three Dimensional Carved Molds

The information above relates to objects with a flat back, such as a pendant or a pin. It is also possible, with a little more work, to create objects with pattern on both sides.

CARVING

Carve the front half of the object as described above. When the recess is completed, brush a thick solution of equal parts of detergent and water into the mold as a separator. Press clay firmly into the carved recess, then use a knife or a thin wire to cut away extra clay, leaving the lump almost flush with the flat face of the mold. Press the second part of the mold against this and carve several notches around the edge for registration of the mold halves. Gently pry the pieces apart so the clay sticks to the uncarved half of the mold.

Use a pencil to trace around the clay, then remove it. You can now carve the second half of the shape and as long as you stay within that outline, the two parts will line up. Casting and venting is as described elsewhere.

Mold Pouring to Make a 3-D Object

Begin by modeling the "front" half of the object in clay. Start by imagining a parting line (the separation between the mold halves) and work up from there. Fix the clay onto a flat surface to secure it as you work.

When the modeling is complete, set the clay onto a piece of glass or plastic and lay strips of wood around it to make a mold frame. Mix plaster and pour it over the clay, allow it to cure for at least several hours, then slide the plaster block off the glass. For registration, cut a couple of notches into the block.

Brush the cavity, the notches, and the face of the plaster with a heavy coating of a 50/50 detergent and water mix, then press a lump of clay onto the first mold section and model the rest of the piece as shown. Wrap a piece of heavy paper or plastic (milk container works well) around the first block and secure it with tape. Be sure this is tight, because you'll have quite a mess if the wrapping falls off. Mix a batch of plaster and pour it over the first mold piece. When it has cured, pry the mold units apart and clean away all the clay, using running water and a soft brush.

Sprues, vents and risers are cut as needed and the casting is accomplished as described elsewhere. Again, remember that the plaster must be fully dry before metal can be poured into it.

INLAYS

Bits of wood, glass, other metals and some stones can be set into a mold to create an inlay.

1. When combining cast pewter with other metals (brass, copper, sterling, etc.) spread a thin layer of flux over the metal to guarantee fusion when the hot pewter flows against it. Use a soft solder flux such a No-Corode (these are typically gooey like petroleum jelly).

2. Allow for a mechanical grip where possible by creating a bevel on the inlaid piece. This should slant so the base, the side in contact with the pewter, is larger than the surface visible in the finished piece.

3. Rough surfaces will make a better grip than smooth ones. Provide a tooth by filing or scratching the surface that will be in contact with the pewter.

4. Small pieces can be glued into the mold with Elmer's or a similar household glue.

This is the section that holds the inlay in place.

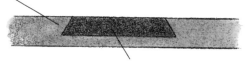

Don't allow this section to become too thin.

Slush Castings

This technique is used to produce hollow castings by interrupting the normal cooling sequence of a thick casting. Start with a mold made in any of the methods described in this chapter, then join the mold halves tightly and pour molten pewter into the cavity until the mold is filled. After allowing a few seconds for the metal to harden, invert the mold and pour out the slushy metal from the inside of the object.

The tricky part is knowing when to pour. If the pour is too soon, the object will have thin areas or holes. If you wait too long the interior of the object will be solid. Consistently good results require some experimentation and a feel for the process, but the results can be worth the time spent.

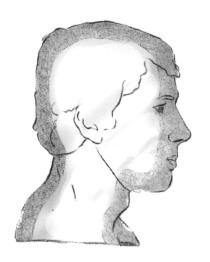

Molds Made of Cast Iron and Bronze

Commercially, pewter is sometimes cast into bronze or cast iron molds. This practice has been eclipsed in recent years by the development of silicone rubber molds which are cheaper to manufacture. The following chapter describes these molds and their use in detail.

Small metal molds can be produced by the ambitious metalsmith in a typical studio. In the case of larger molds, a wooden pattern should be prepared and the casting jobbed out to a foundry. The cost will depend on the metal being used (steel is cheaper than bronze) and the complexity of the mold. Consult a local foundry for estimates.

In use, the mold halves are warmed on a hot plate or in a gas flame to about 200°F/90° C. This is too hot to hold in the hands, so thick gloves are worn to handle the mold. The parts are set together and held in alignment by pins built into the mold. The unit is then clamped together and molten pewter is poured in the gate.

After allowing 30–60 seconds for the pewter to cool, the mold halves are opened and the casting is removed. Because the mold is warmed in the process of casting, it needs to be returned to the flame for only a few seconds to prepare it for the next casting.

CHAPTER 9

FLEXIBLE MOLDS

Introduction

One of the most important developments in casting in this century has been reusable rubber mold. Molds made of flexible rubber conform exactly to the shape and details of an object and allow its perfect replication. More than this, because the rubber can stretch and return to its original shape, rubber molds can create impressions of objects with undercuts. Because of this technology, shapes that previously would have required complex, multi-part molds are now reproduced at a fraction of the cost that would have been incurred only a couple decades ago.

Many kinds of mold materials are available, each with its advantages and drawbacks. Simply put, these usually involve a trade-off between low cost and long life. Other factors that contribute to the decision about which mold material to use are the temperature of the material being injected into the mold and the degree of flexibility required.

Molds made of silicone rubber can withstand temperatures around 600°F/315° C and can be used directly for pewter and other white metals. When casting multiples of an object in brass, sterling or gold, a new investment mold must be made for each casting. Investment molds are by definition waste molds that are destroyed in the process of retrieving the casting. When casting these metals, rubber molds are used to create the wax models that are then set up in an investment mold, burned out and cast in the conventional way.

Room Temperature Vulcanizing Molds (RTV)
(aka *Cold Mold Compounds* (CMC)

One of the simplest forms of rubber molds is made of a polysulfide compound that is purchased as two components, both liquid. When the parts are mixed together, a curing reaction takes place, resulting in a stretchy "vulcanized" rubber. RTV compounds are available from most distributors of casting supplies.

ADVANTAGES
- They require no special equipment.
- Neither heat nor pressure is needed, so almost any model can be used.
- Shrinkage is minimal.

DISADVANTAGES
- The cost is relatively high.
- These molds have a somewhat shorter life than true vulcanized rubber molds. RTV molds are not as tough and are less elastic than vulcanized molds.

← TAB GOES HERE

Process For an RTV Mold

1 Prepare the model by completing it as much as possible.

2 Attach a sprue, using the same logic of positioning as for a standard investment casting (see Chapter 2).

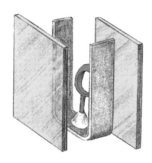

3 Coat the model and sprue with a parting compound such as silicone spray, non-stick cooking spray, debubbleizer or cornstarch.

4 Set the model into a container for mold making. This can be a 3-part rectangular mold as shown here, or something as simple as a paper cup.

5 Mix the two parts of the compound in a disposable container, using a disposable tool. Use the correct proportions, weighing the ingredients if possible to ensure success. If this is not possible, measure by spoonfuls to achieve the proper proportions. It is important that the two components be thoroughly mixed. Use a spatula (tongue depressor, popsicle stick) and work at it for several minutes because an incomplete mix will result in gooey uncured areas within the mold.

6 Some people vacuum the mix at this stage to remove bubbles but it is probably just as effective to rap the container sharply many times. As bubbles appear on the surface, pop them with a needle.

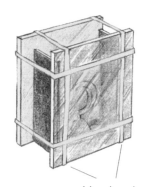

7 Pour the mix over the model, being careful to avoid trapping bubbles. In the case of convoluted surfaces, you may want to begin by spreading a layer of the mix on the model with the spatula or a brush.

rubber bands

8 Set the mold aside to cure before trying to open it. Cure time differs between products, but is often between 16 and 24 hours.

CUTTING A MOLD OPEN

The mold must be cut open to remove the model and allow subsequent wax models to be removed. For flat simple objects, this is fairly straightforward. As the complexity of the form increases, so does the difficulty in mold cutting. Some suppliers stock a translucent mold compound that is especially good for beginners because it allows a view of the object trapped in the rubber.

Use a hook such as a can opener or the bent tines of a fork to help pull the mold apart as it is cut. This "third hand" can be clamped onto the bench or gripped in a vise. Cut the mold with a thin sharp knife. A surgeon's hooked scalpel is best, but any similar blade will do. It is important that the blade be very sharp so plan on using a fresh blade with every mold or two.

Begin at the gate, slicing a line along the bottom edge of the mold about a half inch deep all the way to the corner, then repeat this on the other side. Make a similar cut along both sides of the mold, then connect these two cuts along the top edge. At the bottom edge near the gate, slice all the way into the sprue and from there along the edge of the model. Continue following the outline of the model until the two halves come apart.

Locks

In order to realign the mold halves, locks or registration tabs, are cut into the two pieces. In hand cutting, especially for novices, the cuts are usually irregular enough to provide dramatic contours for fitting the two parts together. With practice the cutting will become smoother and the locks should be consciously included. A zig-zag pattern in the initial cuts makes a good lock. You can also hinge the mold by leaving one edge uncut. For simple molds this can be expedient, but it limits the flexing of the mold and is not suitable for complex models.

RELEASE CUTS

These cuts increase the flexibility of the rubber and are especially useful around delicate, detailed areas of a design. Note that no rubber is removed.

CORES

Some shapes cannot be accommodated by a two part mold and a third part (sometimes more) must be cut out. These can be central units, like the middle area of a ring, or a block that is cut away from the top as shown.

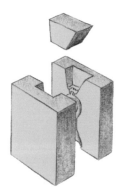

Improvised Rubber Molds

There are a number of silicone-based products that can be used to make flexible molds. The cheapest and most readily available is bathtub caulking, which can be bought at most hardware stores. This cannot be used to make a thick mold, but takes a surface impression nicely. Coat the object being molded with a thick solution of detergent beforehand, and allow it to cure at least 24 hours per 1/4" thickness.

Two-Part Mold Compounds

For greater convenience and results, search out any of the many brands of two-part rubber mold compounds. These were developed for the medical and dental communities and were originally quite expensive, but non-sterile versions are now available at a reasonable cost.

Most products come in two contrasting colors with instructions to knead the putty-like material in the fingers until the colors are completely mixed. Curing times range from 1-15 minutes.

These simple mold materials are mobile, which creates access to a huge range of textures.

Alginate

This dental material, used to make impressions inside the mouth, is made from gelatin-like algae and can be purchased from dental labs or some casting suppliers. It comes as a lightweight flour that is mixed with water to make a mush that is then quickly pressed into place, where it sets up in just a few minutes. The impressions it takes are very exact, but the material is fragile and cannot be flexed without breaking. Alginate becomes brittle within 24 hours, unless it is sealed in a plastic bag, and even then a mold can only be kept for a couple days before it whithers.

When using these impression materials for a flat-backed object, secure the model to a sheet of glass or in the center of a saucer or plastic dish with dab of sticky wax.

Pour/spread the material over the model and allow it to cure. After removing the model, brush wax into the mold—you may find that you need to use a small amount of parting compound to facilitate release. Depending on the complexity of the model, these molds may give from 5 to 50 impressions before they deteriorate.

VULCANIZED MOLDS

I think any caster will agree that the best solution for the duplication of waxes is a bona fide vulcanized rubber mold. Before describing how to make a mold, it should be mentioned that there are many casting companies who will make a mold for you.

All you have to provide is a model, either in metal or in wax. Because wax cannot stand up to the temperatures needed in vulcanizing, a wax model will be cast into metal (for an additional charge), then the company will make a rubber mold of it. The charge for this depends on the complexity and size of the object—a typical ring mold might cost around $40 while the mold for a belt buckle might cost from $70 to $100.

The original model will be returned to you if you request it, but most companies will keep your mold in their plant so they can respond quickly when you order more castings.

VULCANIZING EQUIPMENT

Vulcanizer   This is a machine that combines electrically heated plates with a screw press to convert flaccid raw rubber to a springy material. Machines are available in several sizes and with simple or sophisticated controls. Prices range from $500 to $1000.

Mold frame   This is a thick rectangle of aluminum, commercially available in many sizes. The mold should allow at least a half inch all around the model and about a quarter inch thickness of rubber along the depth of the mold. Frames can also be fabricated, by TIG welding or by bolting heavy bars together.

Rubber   Uncured, unvulcanized rubber is manufactured in sheets 1/8" thick and sold in strips of varying size and rolls 18" wide. The slabs of rubber are protected with a sheet of plastic to keep the surface clean. Remove this only as the frame is being filled.

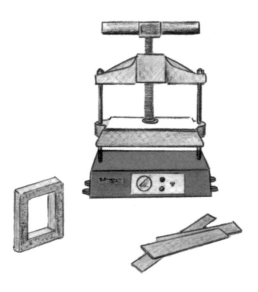

Steps in Making a Vulcanized Rubber Mold

1 Prepare the model as completely as possible. Remember that every detail (including scratches and flaws) will be transferred to the mold.

2 Attach a metal sprue with either hard or soft solder. A 1/8" brass rod is a typical sprue, but many variations are possible. Use the information in Chapter 2 as a guide.

3 If the sprue did not have a cone-shaped button built in (these are commercially available), create a button by fabricating a cone from sheet metal and slide this onto the sprue.

4 With scissors, cut slabs of uncured ("raw") rubber to fit snugly into the mold frame. Supply enough layers to completely fill the frame. For instance a frame that is 1" deep will require 8 pieces of 1/8" rubber. Some companies sell rubber pieces pre-cut to the correct size, in which case no cutting will be needed.

5 Lay half the sheets into the frame, cutting a triangle in the top sheet of this stack to correspond to the sprue base. Lay four registration buttons (steel nuts work well) onto the rubber, then dust the rubber and buttons with talc or cornstarch and set the model into place.

Steps in Making a Vulcanized Rubber Mold *(continued)*

6 Continue building up layers, repeating the triangular cutout in the first sheet. Be certain that each layer of the rubber is clean.

7 Set the frame between two aluminum plates (about 1/2" thick) and set it into the pre-heated vulcanizer, and tighten the plates down lightly.

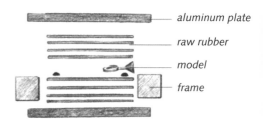

aluminum plate

raw rubber

model

frame

8 After 15 minutes, turn the screw handle to increase pressure on the assembly.

9 After another six minutes, tighten the screw as far as it will go.

10 Vulcanization takes place between 310–325°F/150–163°C and usually requires about 15 minutes per 1/4" of rubber. Test the vulcanization by pulling off a piece of the excess rubber that is squeezed out of the mold through a hole drilled in the frame. If this strand is springy like a rubber band, the vulcanization is complete. If it is brittle, allow the mold to cure longer.

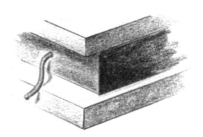

11 When the mold is ready, remove it from the vulcanizer and drop it into water to cool. The mold can be cut immediately.

THE FOUNDRY RUBBER MOLD

Many models can be reproduced in a two-part mold, and that is the example being used here. Advanced casters will be able to make the logical adjustments for complex shapes that require a multi-part mold.

Large scale bronze castings generally use either a blanket or a built-up mold. In a blanket mold, a plaster mother mold is made first. A rubber compound is then poured between the mold and the model to fill in undercuts. In a built-up mold, as explained in detail here, a rubber skin is built up on the model, then supported with a plaster mother mold.

The Process

Determine the location of a parting line, the seam where the mold parts will meet, arranging this to avoid undercuts as much as possible. Roll out a strip of soft clay and lay it into position as a wall along this line. Use a gentle touch and small bits of clay to snug this wall up to the model. It should remain vertical to the surface: don't allow it to lean.

With a clay tool or a loop of wire, cut a gutter in the wall about 1/2" above the surface of the model to provide registration for the two parts of the mold. An alternate method uses small pieces of thin brass or aluminum sheet called shims to form the wall. In this case the gutter is unnecessary because the irregularity of the shim wall provides registration.

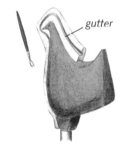

gutter

Mix a workable quantity of an RTV rubber mold compound according to the manufacturer's directions and brush or spread this over the model to a thickness of at least 1/4" using large smooth brushstrokes to avoid trapping air bubbles. The rubber should extend up the wall. Allow the compound to cure as recommended, usually about a half hour. In the case of a simple, relatively smooth-surfaced mold, this rubber coating may be skipped.

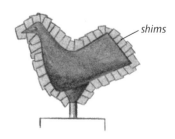

shims

THE FOUNDRY RUBBER MOLD *(continued)*

A plaster mold is then made to contain and support the rubber skin. Mix Plaster of Paris in room temperature water, taking care to create a creamy slurry and a quantity that you can comfortably deal with in 15 minutes. With your fingers or a wide brush, lay the plaster in many coats over the rubber layer to a finished thickness of about an inch. As you near this thickness, lay strips of fabric into the plaster to strengthen the mold and provide hinges to keep pieces aligned if the mold should break. Allow the plaster to cure until it is no longer warm to the touch.

Gently pull off the clay wall around the model to reveal the rubber layer where it ran up along the wall. Coat this liberally with a release agent such as silicone spray, cooking oil or green soap, then complete the mold by repeating the process just described. Brush a 1/8" layer of rubber over the model, including the wall and when this has dried, support it in a mother mold of plaster, again using strips of fabric for additional support. When the plaster has set and the mold halves are pried apart, the rubber will probably remain on the model. Peel this away and immediately set it into the plaster mother mold for safe keeping. Clean any clay or other debris from the inside of the rubber impression.

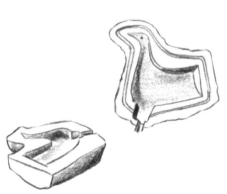

Injection Waxes

Injection waxes are formulated to achieve:

- low viscosity
- toughness
- consistent melting temperature

A number of different brands are available, each with strong advocates—experience will be your best guide. To get started, get the wax suggested by your usual supplier and try it out. Waxes will often by supplied with a recommended injection temperature, generally from 145°–165°F/63–74°C. A usual practice is to try several molds at the lower temperature. If the wax models are incomplete, raise the temperature slightly and try again. Continue in this way until you get consistently sound models.

Wax Injectors

In the case of large, smooth-surfaced models, it is sometimes possible to fill a rubber mold by simply pouring molten wax into it. More often, however, wax must be propelled into the mold in order to reach into every crevice. Hydraulic or air pressure or centrifugal force are commonly used to achieve this.

Wax is melted in a small electric pot that can be controlled by a thermostat to sustain temperatures from 100°–250°F/38–120°C. In the case of a hydraulic pump, a cone-shaped nozzle is attached to a small cylinder similar to a miniature bicycle pump. When the piston is raised the cylinder fills with molten wax. When it is pushed down the wax is squirted out the nozzle. *

A more popular system uses air pressure to blow a charge of wax into the rubber mold. The pressure is pre-set, usually to about 3.5 pounds. Wherever hot wax is being used, caution should be exercised, but this machine is less likely to spray out volumes of burning wax than the hydraulic pot.

A third alternative is to use a centrifugal machine, similar to the kind used for casting metal. Wax is melted in a double boiler arrangement on a hot plate, and a small amount is poured into the crucible of the machine. The spring-loaded machine is then released, causing it to spin and inject the hot wax into the cavity of the rubber mold.

---

* THESE INJECTORS ARE DANGEROUS AND NOT RECOMMENDED FOR CASUAL USE. IT IS EASY TO PUSH TOO HARD ON THE HANDLE AND SPRAY A STREAM OF HOT WAX, RESULTING IN SERIOUS BURNS. EXPERIMENT WITH THIS STYLE OF INJECTOR AT YOUR OWN RISK, AND ONLY WITH GREAT CAUTION.

LUBRICATION

After the rubber mold is complete and the original model is removed, a release agent is used on the rubber mold to assist in the removal of wax patterns. Silicone spray, or cornstarch in a cloth bag are commonly used.

Holding the mold in the hand, flex it to open the cavity and dust or spray the release agent on each half. Too much will cause a loss of detail in the wax model and will also affect the wax if it is put back into the pot. After applying the release agent, tap the mold halves on the table and blow on them to remove any excess.

If the wax is sticking to the mold even when release agents are used, the problem may be that the wax is too hot. Adjust the temperature so it is just barely above the melting point of the wax.

CHAPTER 10

FOUNDRY CASTING

Introduction

There is not a specific definition for "foundry casting" except to say that it describes all work done in a foundry. For the purpose of this chapter, we'll use the term to mean the casting of hollow objects that are too large to be cast in a jewelry studio, say between 8" and 30." Of course much larger pieces can be cast in commercial foundries.

Many of the techniques described here can be used on both larger and smaller castings and can be used with bronze, aluminum, brass, and in some cases, cast iron. For simplicity I will restrict the text to bronze, but those with interest in other alloys should experiment carefully and pursue further reading. A visit to a local foundry is also recommended as a fascinating and instructive experience. This chapter is presented to pique interest in this aspect of metal casting.

OVERVIEW OF THE PROCESS

1   Make a hollow wax model.

2   Attach gates, risers, vents, core pins, etc.

3   Invest the model and fill the core.

4   Burnout.

5   Melt and pour the metal.

6   Weld pieces together, make repairs.

7   Chase the casting.

8   Clean and if desired, patina the object.

Ideally these steps flow together and relate to each other. Each design brings its own challenges and peculiarities, so the best that can be done here is to give a generalized sequence.

We are assuming here a model made in clay, plaster or wood. When the sculptor has finished his or her work, the founder takes over. Of course it might be that the artist and founder are the same person.

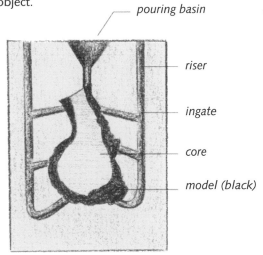

pouring basin

riser

ingate

core

model (black)

TAB GOES HERE

Gating

Determining the number, length and location of the channels that will bring molten metal into the mold is a complicated business and in the end, a largely subjective one. What follows are some common systems with the reasons behind them.

The central concern of every component of the gate system is to provide for the efficient, non-turbulent flow of metal into the mold cavity.

To achieve this:

1 The wax rods should be smooth and of consistent diameter, in progressively graded sizes (large, medium and small sprues, depending on how much metal each needs to carry).
2 The joints between rods should be secure, smooth and lightly filleted.
3 Ingates should be slightly tapered toward the model.
4 Secondary gates always run uphill, to fill the model in sequence.

Glossary of Gating Terms

GATE   A funnel-shaped entrance to the sprue or runner. (aka Pouring Basin and Cup)

GATING   The science and art of placing sprues, runners, risers and vents.

INGATE   The point where a sprue attaches to a model.

RISERS   Openings (thick vents) in a mold that provide reservoirs of metal whose delayed cooling compensates for shrinkage within the casting. These are sometimes called feeders.

SPRUES   The channels through which metal is fed into the mold. A large central sprue is called a runner.

VENTS   Wax rods (later tunnels) that assist in the removal of molten wax and release air from the mold as hot metal enters. Also called whistlers, because of the noise made as air is pushed out by molten metal entering at the other end of the mold.

Determining Gate Locations

- *Metal shrinks as it cools.*
- *Thin sections cool faster than thick ones.*

These simple statements are the basis for sprue arrangements. Lay out a system of feeders that provides for the thinnest section of the model to harden first. It should be fed by a thicker section which will in turn should be fed by the thickest section of the model, which is in turn supplied with molten metal from an even heavier section (riser, runner, or sprue) outside the model. Because this will be the last area to chill, this is the place where porosity will occur.

Avoid having a long drop of molten metal in the mold cavity be-cause the weight of the metal might break off chunks of the mold. Also the metal may splash about as it hardens, causing a loss of detail on the surface. It is generally better to feed the casting from the bottom - picture filling a teapot through the spout.

It should be obvious from the above information that a lot of thought goes into devising a gating system. And remember, this is sim-plified information. Use sketches to methodically trace the flow directions of pouring and hardening, then check and double-check your plans.

Making the Wax Model

1   The first step is to create a model that is the exact shape and size of the desired object. For this example, let's assume it is made of clay. In order to duplicate every detail of the original, a rubber mold is made as described in the previous chapter.

2   In a double boiler arrangement (see Chapter 1), heat wax until it is above solidus temperature. With a soft brush, paint a layer of wax on the inside of the rubber skin in both parts of the mold, giving special attention to details and heavily contoured sections of the design. The wax should be at its most fluid so it will fill every detail of the impression.

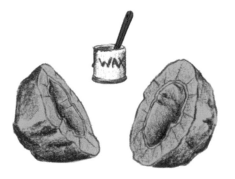

3   Areas with sharp angles should be built up with a wax that melts at a higher temperature because these areas might suffer erosion during the next step, when molten wax might wash this first coating away.

4   Set the two halves of the mold together and tie them with a stout cord—use wedges or a tourniquet to keep the parts snugly joined. Heat a bucket of wax to a point just above solidus, that is, almost ready to harden and pour it into the mold.

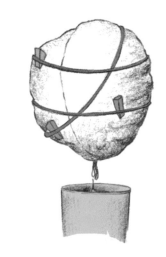

   The goal is to create a wax skin that is the same thickness as desired for the bronze casting. For most objects this will be between 1/8" and 1/4." Roll or tip the mold around to be sure that the interior is evenly coated in a technique similar to the ceramic process called slip casting. After a few seconds, pour the excess wax back into its bucket.

Making the Wax Model *(continued)*

5   The wax must be completely hardened before opening the mold. If you're in a hurry, this can be hastened by filling the mold cavity with cold water. Pry the plaster molds away and gently peel back the rubber. The result should be a hollow wax model that perfectly duplicates the original clay model. It is not unusual for small blisters or craters to occur, but these can be easily rubbed off or filled with a soft wax. Of course if the wax is severely malformed, clean the wax, melt it down and try a new pour.

- Some models are best handled by cutting off sections, casting them separately and welding them back onto the object. For those cases, this is the time when the wax model is cut up. If the piece is to be cast whole, this is the time when core windows (access ports for removing the core) should be made.

- To safely store wax, float it in a tub of cool water, or suspend it in a hammock of string. Note that the rubber impression and plaster mold are intact and can be used to make many subsequent waxes.

Investing

1 To contain the mold, make a cylinder of plasterer's lath, chicken wire or a similar wire mesh. Plan a size that allows about 2" around the sprued model and close the cylinder with twists of wire.

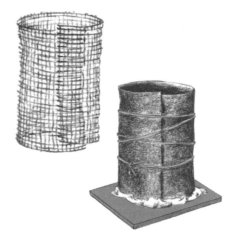

2 Wrap the cylinder with roofing paper (tarpaper) allowing several inches of overlap, and secure the it with a stout rope.

3 Set the cylinder on a stable flat area (e.g. plywood) and splash a layer of plaster along the bottom, inside and out, to seal the cylinder.

4 The surface of the wax model is now coated with investment and because this is the surface against which the metal will form, it must be exact. Use a good quality investment and pay attention to proper mixing. See Chapter 3 for a detailed description of this material and its use. It will remind you to wear a respirator.

5 Fill the cylinder 3/4ths full with investment, then gently dip the stuccoed model into this investment until the cup is flush with the top of the cylinder, allowing time for the investment to enter the core. To facilitate this, all openings to the core should point upwards. Hold the model securely, being sure it doesn't touch the sides of the cylinder.

Investing *(continued)*

6 Add more investment to fill up ("top off") the cylinder. Either hold the model until the investment has set up or arrange some way of keeping it suspended in the correct position. If you tie it, be sure that it doesn't swing sideways and come in contact with the wall of the cylinder.

Where economy is important, use a backup investment made of one-third plaster of Paris and two-thirds refractory material.
These could be:
    sand
    vermiculite
    grog
    used investment (ludo)
    cristobalite
    silica flour (wear respirator)

Whichever you use, add about 10% wood flour (fine sawdust) to the mixture, for permeability.

7 Allow the mold to dry completely before going on to the next step. In a warm dry environment this could happen in a few days; in cool damp climates the drying could take as long as two weeks. To hasten the drying, peel off the paper shell when the plaster has set.

Burnout

The logic and technique of burnout for foundry casting is the same as that described earlier for lost wax investment casting. The only difference is in size and time. The purpose of burnout is to heat the mold to remove the wax model and gates, to clean the mold cavity, and to cure the investment. All this is accomplished by slow heating to 1250°F/675°C. If a gas kiln that can accommodate the mold is available, simply set the mold into place with the sprue cup facing down, and let the kiln do its work.

Because the scale of foundry casting uses a good bit more wax than jewelry casting it makes a lot of nasty air, smelly and noxious. Ventilation is necessary.

HOLDING THE CORE

When the wax model has been burned out, the core will fall unless provision has been made to keep it in place. In some cases the core may be bridged into the surrounding mold because of the shape of the object. More likely, the core will make connection at some place, such as in the neck of a bust. This helps, but is not sufficient by itself. Provide further support for the core by using core pins or chaplets as shown.

Locate the chaplets in positions that can be easily reached and repaired, preferably areas that will be ground smooth or left heavily textured—subtle textures are the most difficult to repair.

Chaplets are usually slid into the wax model before filling the core. For most castings, use galvanized nails for chaplets. These make a good grip in the investment because of their rough texture and because the hot metal burns the zinc coating away removal is relatively easy. For small castings, stainless steel sewing pins can be used.

It is also possible to use rods of the metal being cast. After casting, excess is cut off and the stub is chased flush with the surface. This is especially recommended for textured areas.

When the casting is completed the chaplets are pulled out or snipped off and driven into the core area, from which they can be shaken out. The holes that are left will need to be patched, which is explained below.

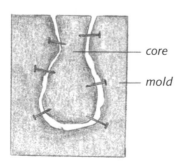

core

mold

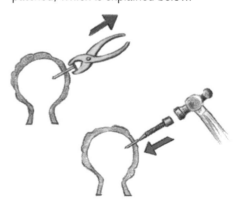

WAX TOOLS

Chapter 1 describes electric wax pens, flame-heated needles and soldering irons used to weld wax pieces together.

Weigh the wax model with its gating system to determine the amount of metal that will be needed for the casting. Write this down, and later mark it on the investment mold. For bronze you will need 10 times the weight of the wax.

Burners

For small kilns, it's possible to use a torch as a burner by anchoring it in a cradle of bricks. It will be necessary to leave several inches of air space between the kiln and the burner to allow proper combustion. The correct flame will be indicated by a low steady roar. There should be no odor of fuel.

Inexpensive burners can be bought from suppliers of ceramic equipment. Because the wax residue from burnout can contaminate a kiln, your neighborhood potter won't want you using his or her kiln, but you may be able to borrow a burner.

If you have any questions or reservations about the installation of a burner, ask the local gas company to check it out before you light it.

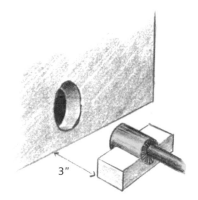

3"

Lighting a Furnace

1   All materials—bricks, mortar and mold—must be dry. Keep the lid off while lighting the burner.

2   Test all fuel connections by brushing them with soapy water. If there are any bubbles, tighten and recheck. Use Teflon pipe tape on all joints.

3   Set the crucible in place, loosely charged with bronze pieces, and place it on a layer of cardboard to prevent its sticking to the base of the kiln.

4   Set a wad of burlap or paper into place well in front of the burner.

5   Light the wad and turn on the air about half strength. Get it burning briskly.

6   Turn on the gas and adjust to a low roar. Stay with the furnace, ready to turn off the gas if the flame goes out.

7   Allow about 5 minutes for the furnace to come up to a temperature that will sustain the flame, then turn up the fuel to full throttle and adjust air to maximum roar. Open the air valve a little more (or cut back on the fuel) so as to make an oxidizing (pale, hissing) flame. Set the lid into position.

Building a Kiln

Because of the large size of the mold, it might be necessary to build a burn-out kiln, which can be done in several ways. In addition to these drawings, consult a potter or ceramic supply company.

One solution involves building a temporary structure of firebricks around the mold. This is best done outdoors, at least 10 feet from any combustible surface. Set the mold on a platform of bricks and arrange the structure with several holes or ports. One is needed to insert the burner, another is needed to view the mold, and at least one more should be included, in or near the top, for a vent.

As shown, the burner flame is directed at a tangent to the mold, so it spirals around the cylinder to create even heating. Bricks set diagonally into the corners will help steer the heat.

The lid can be made by bolting firebricks together, or by fixing them into a steel frame. An alternate method is to build a frame and fill it with a castable refractory cement, transite, firebricks or Fiberfax.

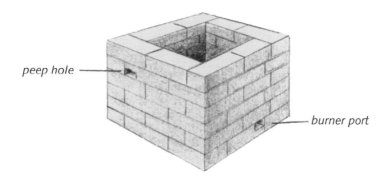

peep hole

burner port

top view

Building a Drum Kiln

A more permanent kiln can be made in a drum, like this. Use #20 insulating bricks or a layer of Fiberfax blanket to line the inside of a clean oil drum or trash can—the size you use will be determined by the largest mold you foresee burning out. The lining will need to be about 2" thick, and the mold should have clearance of 3" all around.

This construction requires the same three ports mentioned above: burner entrance, viewing hole, and vent at the top.

Make tie-downs from stainless steel and nichrome wire.

Burnout

Plan the kiln so that the pouring cup or a riser is visible. When burnout is complete, these openings will be clean and white. If these are all facing down and therefore out of sight, allow the mold to bake until there is no smell of wax vapors.

Large molds are usually burned out with the pouring basin on top. When done this way, a drain hole must be provided on the bottom for the wax. When burnout is complete the hole is plugged with fireclay, the furnace is removed or dismantled and the casting can be poured. Smaller molds are removed from the kiln and set into a casting pit, where several molds might be filled in a single pour.

The casting pit is a hole in the ground or a large box filled with loose dirt that secures the molds in position. Ram the dirt around each mold firmly to fill in any cracks in the plaster mold.

Melting the Charge

MELTING METALS PRODUCES FUMES THAT ARE HEALTH HAZARDS. WHEN WORKING AROUND MOLTEN METAL, ALWAYS WEAR A RESPIRATOR, HEAVY GLOVES AND PROTECTIVE CLOTHING. EYE PROTECTION, ESPECIALLY OSHA-APPROVED DARK LENSES ARE ALSO A GOOD IDEA.

Before you begin this step, it might be wise to make a dry run of the casting process to insure that the pouring ring is the correct size and in a handy location. A little rehearsal never hurts.

It is possible to melt metal in the same furnace used for burnout, but this is not recommended. The burnout kiln is used to slowly take a large vessel to a temperature of 1250°F/675°C. The melting furnace is designed to heat a smaller vessel to a higher temperature as quickly as possible. By definition the same furnace cannot accomplish both these tasks well.

Melting will take place in a barrel-shaped crucible made of graphite (plumbago) or silicon carbide. Consult online resources for a list of the available sizes and the quantities of metal they hold. Use the instructions above to construct a furnace around the crucible, leaving 2-3" clearance all around.

Again, remember to direct the flame on a tangent to the crucible so as to force the flame into a spiral ascent to prevent hot spots. The furnace can be a simple structure of unmortared firebricks but it should have a lid with an exhaust hole.

Venting the Core

As molten metal is poured into the mold, gases in the core will need to escape. If no vent is provided for them, they will bubble out through the bronze, causing a rough and possibly flawed casting.

If a large piece of the core is exposed, as in the neck of a bust, gases can be vented through a tunnel running into the core that is made by wrapping a dowel with waxed paper and sliding it into the investment before it hardens. This dowel will be removed before the mold is burned out, but in the meantime it can provide a useful handle, at least for a small model.

Gases and Oxides

As brass and bronze melt they tend to dissolve gases that will bubble out of the metal as it is poured and can ruin a casting. Hydrogen and carbon monoxide are frequent sources of the problem. To deal with this, chemicals can be introduced into the melt to scavenge out the unwanted gases. These chemicals usually bring oxygen into the melt, where it combines with the gases to create compounds (like water) that are easily driven away.

Sometimes the oxygen intended to purify the metal forms oxides, a further contaminate. A popular solution to this problem is to plunge pellets of phosphor copper alloy into the melt. The phosphor eagerly combines with the oxides and goes off as a gas, while the copper simply goes into solution.

Suppliers of foundry metals are your best source for detailed advice on what compounds and quantities to use in your situation. When speaking with them, you will need to know the size of the melt (in pounds) and the specific alloy you are melting.

TOXIC GASES ARE PRODUCED SO WEAR A
RESPIRATOR AND PROTECTIVE CLOTHING.

Pouring

Test the fluidity of the melt by stirring the metal with a preheated carbon or iron rod. If you can feel no chunks, the metal is molten but probably not quite ready. Most bronze alloys melt around 1900°F/1040°C and should be poured at about 2150°F/1175°C. At the same time, don't overheat the metal, because this will encourage oxidation and shrinkage.

Have the mold rammed up ready so you can pour as soon as the melting is complete. A person working alone can pour up to 20 pounds but it is difficult and not recommended. This is a good time to have a friend.

1   Lift the crucible with tongs and set it on a brick positioned inside a pouring ring. Moving with efficiency, but not a frantic haste, position yourselves above the mold and tip the pot slightly. Use a preheated steel rod to skim dross from the surface of the molten metal.

2   Pour the metal into the funnel-shaped basin, trying to sustain a pace that will keep the basin filled right to its top (this helps prevent dross from entering the mold cavity). Continue to pour until metal is seen at the top of the risers and vents. Do not interrupt the flow of the molten metal.

3   If there is extra metal in the crucible, pour it into the adjacent dirt or an ingot mold. If allowed to cool in the crucible, it will certainly be difficult to remove later, and could break the crucible. While equipment is still hot, use a steel bar to scrape the inside of the hot crucible clean, removing any flux and dross that sticks to the sides, then set the crucible back in the kiln, now turned off, to cool slowly.

ingot mold made from angle iron

Shake Out

The mold should be left undisturbed until the metal has solidified completely. The time needed for this will vary greatly, from a half hour to overnight. When the top surface of the gate changes from shiny to matte, and feels solid when tapped, it's safe to break into the mold. Be patient: slow cooling leaves the casting dense and malleable—both desirable qualities.

Use a mallet and a dull chisel to break off the investment but be cautious of burns; it's still hot in there. In delicate areas, substitute a piece of wooden dowel for the chisel, always remembering that it's better to be cautious and take your time than to risk damaging the casting.

The sprues and runners are cut off with a hacksaw or bolt cutters, with care taken that you don't cut too close to the casting. Avoid rocking the tool back and forth because this can distort the shape of the casting.

When most of the investment has been removed, the cast is washed or hosed off and scoured with a stiff brass brush. Beadblasting or sandblasting also work well as cleaning processes.

Pull out the chaplets with tongs or a heavy wire cutter or, where they cannot be withdrawn, cut them off and tap them into the casting with a punch and hammer. These holes will be repaired later.

Remove the core by breaking it into small pieces or powder. Obviously castings with a small entrance to the core make this job more difficult. Chasing will often shake out stubborn last bits.

Pickling

After casting, the bronze might have an unattractive dark and mottled appearance which can be removed chemically in an acid called a pickle. Dilute Sparex or formic acid are recommended. Wear a respirator and eye protection.

The length of time in the pickle depends on the alloy and the thickness of the oxides, but could be from fifteen minutes to several hours. When the metal is pretty much a uniform color, remove the casting and rinse it well in running water. Even after a thorough flushing with water, traces of the pickling acid will remain in pores in the bronze. An effort to remove them with heat only drives off the water, leaving a salt that will eventually recombine with moisture in the air and reconstitute the acid.

To beat this problem, neutralize the acid in a basic solution such as a concentrated mixture of baking soda in water, a box to a bucket. Allow the casting to soak in this solution until it is completely permeated, like overnight. Rinse the casting off again in running water and use a blast of air or roll it in sawdust to dry it. This is highly recommended because passive air drying or the use of heat might result in a mottled surface, which is what you were trying to remove in the first place.

Repairs

When it comes to patching the holes and flaws in a casting, each individual has his or her private bag of tricks. Here a few of the typical methods, but bear in mind that this is an overview. You should devise your own methods and solutions as needs arise.

COMPRESSION

To pack down the relatively loose structure of a pitted section, grind the teeth off a ball bur, leaving a pattern of random facets, that are then sanded and polished. When this tool is set into a drill or flexible shaft machine and spun against the casting, the facets act as miniature hammers that pummel the surface and compress the pits. A similar effect can be achieved by filing the end of a rod into a blunt hemisphere, then bending it slightly and polishing it smooth. It is used in the same way, with the bent tip of the rod pounding the metal with each rotation.

WELDING

Bronzes can be patched by welding (fusing) pieces of the same metal in place. Pieces of sprues and runners are used for this to guarantee color match. A TIG welder is best but an oxygen torch will do the job. Wear dark goggles.

Borax may be used as a flux. Heat the tip of a rod and dip it into powdered borax which will adhere to the rod. The fumes created in this process should be vented.

A solid weld requires that all the sections involved are at their melting point. Failure to achieve this temperature can make a cold joint, one that looks solid but is fragile. Test your joints as you go along to be certain they are sound.

SILVER SOLDER

This is like welding, but takes place at a temperature well below the melting point of the bronze which makes it a little safer for the inexperienced welder. The principle disadvantage is that there will be a color difference in the seam.

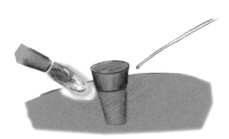

TAPPING

Holes can be threaded with a tap and plugged with a threaded rod made from a sprue. After screwing the plug in securely, cut it slightly above the surface and tap it down with a matting tool to flare the edge and lock the plug in place. Failure to do this might allow the plug to vibrate loose someday.

Use a tap with a screw-unscrew motion to create a threaded hole.

Use a die to cut matching threads on a wire of the same metal (often a sprue from the same casting). After screwing it into place, trim and planish to lock the plug in place.

MECHANICAL INLAY

Use a bur to open out a hole with an undercut like this. Make a plug of the same metal as the casting and set it into place, then tap it with a small hammer until it flares out enough to fill up the hole. This is easiest when the inside area can be supported on a stake or similar metal support while hammering.

The intended result.

PLASTICS

A number of resins can be used to make cold patches. Auto body filler is a good choice, and easily obtainable through local supply companies. This can be used as is and colored with artists' oil paints after it hardens. Experimentation will be required to match the color of the patina—it will probably involve painting several colors in successive layers.

Plastic resin can be mixed with bronze filings to make a paste called cold cast bronze that will color after burnishing somewhat like solid bronze, but again some touch-up painting might be needed.

Most of the information in the preceding chapters has fallen neatly into place. But like everything else in life, casting has loose ends that don't fit quite as nicely. That's why there are last chapters.

Most of the information here can be described as refinements or variations on the techniques already described. They are not magical, and are usually the direct extension of a pretty simple idea. Keep in mind that (a) these are only a few of the many variations that have been invented so far, and (b) the inventing ain't over yet.

Double Metal Casting

This trick is used to cast a single object in more than one metal. It's tempting to think you could throw two metals into the crucible, melt them together, and get a bimetal casting but you won't. You'll get a solid casting in a new alloy. This may be an interesting metal, but the odds are against it. Metallurgists have worked literally for centuries to come up with alloys that combine ideal working properties with attractive finishes. The results of those labors are the alloys we usually use—sterling, karat gold and so on.

Double metal casting is a two-step process—really a pair of castings worked one on top of the other. The example described here uses two metals, but of course three or more could also be used. Each metal is cast in a separate operation.

TAB GOES HERE

PROCESS

1 Make a wax model for the first piece. Depending on the shapes involved, you might carve/sculpt this all by itself, or you might create the entire shape and then bisect it.

2 Cast the first unit using any method you prefer. Cut off the sprues, pickle it, and finish it through the medium abrasive paper stage.

3 Shape the rest of the wax model onto the first piece, using either modeling or carving waxes as called for in the design. Though not absolutely necessary, the results are much better if you can arrange a mechanical grip between the two units.

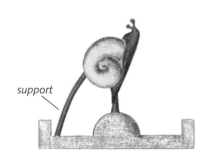

example of a mechanical bond

4 Sprue the wax area and set the model onto a sprue base (it may be necessary to support the model because of the weight of the first section). Notice that because this support does not connect with the sprue it will not feed metal into the mold cavity.

5 Burnout and cast as usual.

support

After casting, the two sections usually seem to be joined but this can be deceptive. Because of the oxide layer formed on the first unit during burnout, the interface between the two metals is probably not fused together. For this reason try to insure a mechanical bond between the first and second castings. At the very least these will hold the pieces tight enough to keep them together while the casting is cleaned of investment. Pickle and neutralize the surface, then flux and solder the joint, before subsequent filing and sanding risks distorting the perfect fit achieved in casting.

Casting Tubes

Tubes are cast to make hinges, ferrules, findings and settings. There is a potential problem because if investment is blocked from the tube by the air already inside it, the casting will result in a solid rod instead of a tube. A core is needed.

For short tubes, use a brush or miniature spatula to pack investment into a model. If one or two tubes are involved, there is time to do this during the usual investing process. If many tubes need to be packed, mix a small amount of investment and pack the tubes first, then invest the model as usual.

Because graphite withstands high temperatures and does not stick to molten metal, it makes terrific core material. Pencil leads provide a convenient source for thin graphite rods—they can be bought through art and office supply stores in a range of sizes, and of course you can simply cut or burn the wood off a pencil. Soft grades are preferred because they are easier to remove. Graphite can be filed or sanded as needed if the rods you buy are too large.

Form a wax shell around the graphite rod, allowing several millimeters to extend out each end. This area will be gripped in the investment around the mold cavity and will hold the core in place when the wax is burned out. After casting, the graphite is broken out with a wire or drill bit.

When using a rubber mold for objects with tubes, set a graphite rod into place in the model before making the mold. This will create a channel for the core in the rubber. Before injecting wax, lay a piece of graphite into the channel so the wax will flow around the core and come out of the rubber mold ready to cast.

Making Hollow Models

Hollow objects can be made by carving the completed form, then splitting it in half and hollowing out the parts. These can then be invested and cast as two separate objects that are soldered together later. When soldering, provide an escape vent for the air caught in the form. Failure to this can cause an explosion as the trapped air expands.

It is also possible to split and hollow a wax model, then rejoin the wax pieces. Hollow models can also be fabricated from wax sheet and wires.

CASTING A CUP

This example describes a shape that is hollow but provides access to the interior of the form, a situation possible in vessels, handles, short spouts and ferrules.

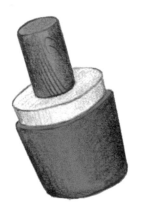

Make a core from investment either by pouring investment into a clay mold or by carving a previously hardened block of investment. Remember that the core must be smaller than the desired shape by an amount equal to the thickness of the final casting. It is sometimes helpful to fix a dowel in the core to make handling easier.

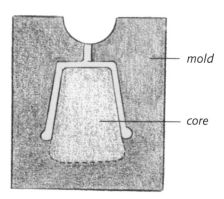

mold

core

HOLDING THE CORE

In cases where the core is mostly enclosed in wax, steps must be taken to hold it in place after the wax has been burned out or else the unsupported core might shift. This is done by placing pins (called *chaplets*) through the model. Use small galvanized nails or short straight lengths of iron binding wire, usually slid into place through the wax. If they cannot be pushed in without distorting the wax and core, set them in a drill and "screw" them in.

After casting and before pickling, pull these out with pliers. If they cannot be withdrawn, cut them off flush and use a scribe to drive them into the core area so they can later be shaken out and discarded.

Pins will of course result in holes in the casting that will have to be plugged later. After pickling, use a round file to clean the edges of the holes and push tapered rods of the metal used for the casting into place. Use silver or gold solder as appropriate for precious metals. Bronze should be welded, but silver solder can be used in a pinch. Do not use a tin- or lead-based solder, because subsequent repairs might result in contamination of the metal.

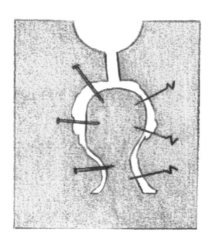

Nesting Shapes

A popular application of this effect is in wedding ring sets, where the engagement and the wedding band are shaped to fit snugly against one another. This example will use a wedding set, but many other applications come to mind.

It is possible to simply carve two wax models that fit together, a task that is easy enough if the interface between the two is perfectly flat. In cases where the contours are flowing or involve complicated facets, it becomes difficult to make a perfect fit.

---

1   Carve, invest and cast one of the rings.

2   Finish it, at least through a fine sandpaper finish.

3   Rough out a block of wax from which the second ring will be carved. In the case of modeling (soft) wax, create the necessary form by building up layers of wire and/or sheet and skip the next step.

4   Hold the finished ring in tweezers and warm it slightly in a spirit lamp or a gentle torch flame. Press the cast piece into the wax, being careful that it doesn't go in too far.

5   Allow the wax to cool , then carve the second model with the metal mate still in place. It is possible that the two parts will separate during the carving process—just slide them back together and continue. If they haven't come apart by the time you're done, simply pry the two units apart.

6   Sprue, invest and cast the second piece. Because of shrinkage (2-5%) the fit will require some fine tuning.

Implants

Implanting is similar to the double casting just discussed, but usually implants are relatively small details added to a design. Implanting is done:

- to achieve an integral look between components.
- to incorporate mixed media, for instance as an inlay of a contrasting metal.
- to create in metal those shapes that are too frail to be constructed in wax.

HOLDING AN IMPLANT

The implant must be securely held in place, even after the wax has burned out. This drawing shows what will happen if the implant is free to drop away from its intended position. Obviously, the result would be a ruined casting.

Investment is somewhat adhesive and may grab onto the implant tight enough that it will stay where you want it, but it is unwise to depend on this. Be sure that a curl of investment grabs onto the implant. Common sense and some experience in visualizing the interior of a mold are needed to apply this rule to a variety of situations.

It's sometimes necessary to solder or in some other way provide a hook for each implant. These are then cut off after casting. Notice how pairs of implants are shown using the same hook (in this case a wire soldered between the two). Because the interior of the mold is never heated above 1250°F/675°C, any grade of silver solder will suffice. Do not use a lead- or tin-based solder! It will contaminate the silver or gold.

BEZELS

The narrow rim of metal that is pressed down over the edge of a stone to hold it in place is called a bezel. These are usually quite thin (about 28 gauge B&S) and wax this thin is so fragile that even a tiny slip with a hot needle will distort it—in fact even holding it lightly in the fingers can twist it badly out of shape. For this reason it's often easier to make a bezel of silver or gold and connect it to a cast piece. The bezel can be soldered to a finished casting, but it is also possible to cast the bezel in place.

To implant a metal bezel, start by fabricating a bezel in the desired metal. After checking for the correct size and shape, press it into the wax model and drip soft wax into place around the bezel as shown. Be sure to add wax only on the outside of the bezel or you will destroy the fit. If the model provides adequate support for the stone, a simple "bottomless" bezel is usually all that has to be fabricated. In other cases, the bezel rim should be soldered to a base of sheet metal, which should have a couple small holes drilled through it to create a mechanical grip.

Gems as Implants

Some gemstones can survive the heat and thermal shock of being cast in place. Remember though, that for every gem that has been cast in place, about a billion have been set with mechanical means (bezels, prongs, etc.). This is one of those cases when the weight of tradition is worth considering.

Stones cast in place tend to sink into the design, like a pebble in mud. There are times when this effect can be used to advantage, but those are special cases. Be sure in your own mind that the reason for casting the gem in place is because of design and not an alternative to proper setting in the traditionally accepted manner.

Not all stones can be cast in place. The "earthy stones" such as agates, turquoise, malachite, lapis, etc. will almost certainly lose their color and crack. Organic materials like amber, coral and ivory are definitely doomed to failure. Clear hard stones such as diamond, ruby, sapphire and emerald will probably survive the burnout and casting operation, but beware because impurities in the stones can result in cracking. Synthetic stones such as cubic zirconia, whch are often grown in high temperatures are likely to survive the process.

STEPS FOR GEM IMPLANTS

1 Hold the stone in tweezers and clean it with alcohol to remove finger oils.

2 Carve a hole in the wax roughly the size of the stone, then warm the gem slightly and set it into place. Be careful not to overheat the stone—this will cause it to sink too far into the wax which is a messy nuisance to correct.

3 Provide for a "finger" of investment to hold the stone such as a hole in the wax (for instance an opening between prongs). In the case of irregular stones such as crystals, this may not be necessary because the investment grips the rough surface of the gem.

investment will flow around the stone here, holding it in place

4 Complete the model. Invest, dry, burnout and cast as usual.

5 Allow the mold to air cool after casting to reduce the thermal shock to the stone. When the mold is cool enough to hold in the hand, scrape away investment to retrieve the casting.

Alcohol Lamp	a small lamp, usuailly a bottle with a wick, used to provide a small clean flame when working with wax. Also called a spirit lamp.
Alginate	a semi-flexible mold material, popularly used in the dental industry. It is made from seaweed algae.
Borax	hydrated sodium borate, a crystalline compound that is mined, refined and used as flux.
Bottom-Pouring Ladle	a steel ladle used to pour low-melting alloys like white metal and pewter. The bowl of the ladel is divided by a wall that has a hole or holes at the bottom.
Burnout	the processs of removing a wax model from an invest ment mold by heating the mold in a kiln or furnace.
Button	the lump of metal that fills the sprue base in a jewelry casting.
Chaplets	pins used to anchor a core in place.
Chasing	a technique that uses stamps to add detail to a casting.
Cope	the top half of a casting frame used in sand casting.
Core	a section of a mold that corresponds to a hollow interior of the final casting.
Cradle	a sheet of metal hung over the arms of a centrifugal casting machine to elevate the flask so it aligns with the crucible.
Crucible	a vessel of refractory material such as ceramic, graphite or silicon carbide.
Cup	in foundry casting, the entrance for molten metal.
Cutoff Needle	a pointed steel wire in a simple wooden handle. Also called a biology needle and a needle tool.
Cuttlefish	a small squid-like animal whose soft porous skeleton can be used as a mold.
Debubblizer	a wetting agent used to facilitate the coating of a wax model by investment.
Double Metal Casting	a technique in which two metals are cast one upon the other.
Draft (Taper)	the slanted side required for a model to be sand cast.
Drag	the bottom section of a sand casting frame

↑ TAB GOES HERE

Dross	the oxidized waste created when melting low temperature alloys.
Extruder	a tool that forces softened wax through a die to create specific shapes.
Expanded Plastics	light-weight plastics such as Styrofoam and styrene. These can be used as models.
Fillet	a small bit of material added to a joint to enlarge it at the point of contact.
Firebrick	refractory bricks made specifically for use in high temperature applications.
Flashing	a flaw that creates "feathers" growing on the surface of a casting.
Flashpoint	the temperature at which wax suddenly ignites. The exact temperaure varies with each wax.
Flask	a vessel or form used to contain and support a mold.
Flux	a chemical used to absorb oxides and promote melting.
Foundry Sand	finely ground sand used to make molds.
French Sand	one kind of foundry sand. Also called Green Sand.
Gate	the entrance to a mold into which molten metal is poured.
Hard Core (Shell Casting)	a method in which a relatively thin layer of investment is coated onto a model.
In-gate	a secondary sprue that feeds inrushing molten metal from a main sprue to the model.
Investment	a plaster-like material made of gypsum, silica, and cristobalite.
Kiln	a high temperature oven used to burn out a mold.
Model(Pattern)	an object that creates a negative shape in a mold. The model is the same size and shape as the desired casting.
Mold (Mould)	a negative impression of a desired shape. It is in this that the casting is made.
Mold Frame	a thick aluminum frame in which rubber is vulcanized.
Molding Boards (Moldboards)	panels of wood slightly larger than casting frames, used as lids and bases for the mold in sand casting

Mother Mold	an exterior supporting mold.
Needle Tool	another name for a cutoff needle.
Oxidizing	the process in which oxygen combines with elements to create new compounds called oxides. These are something to avoid.
Pattern	another name for a model.
Parting Agent	a substance used to prevent parts from sticking together.
Parting Line	the point at which mold pieces separate.
Perforated Flask	a flask with many holes used in vacuum-assist casting.
Piece Mold	a mold that comes apart in two or more pieces to allow for removal of a model.
Pits	tiny holes in a casting. These can be caused by oxides, impurities in the metal, and by contraction of the metal.
Pounce	a material such as cornstarch or talc used as a parting agent.
Pouring Basin	in foundry casting, the entrance for molten metal.
Pouring Crucible	a ceramic ladle used to melt metal and pour it into a mold.
Recrystalization	the process by which a metal changes from a liquid to a solid state. Also called "freezing."
Reducing Flame	a fuel-rich flame that helps prevent oxidation.
Registration	the alignment of parts.
Release cuts	slits in a rubber mold that provide additional freedom of movement. Some molds need release cuts to bend sufficiently to remove a wax model.
Riddle	a mesh used to sift sand in a process called "riddling."
Risers (Vents)	components of a sprue system designed to allow gases to exit as metal enters a mold.
Room Temperature Vulcanizing (RTV)	a family of flexible mold compounds that cure without the use of heat. Also call "cold mold compounds.
Silicone Rubber	a flexible material capable of withstanding high temperatures, used to create a seal in vacuum casting, and for direct injection molds for white metal alloys.
Silicone Spray	a parting agent

Slush Casting — a technique used to make hollow castings. A liquid material is poured into a mold and poured out again as it begins to harden. In ceramics this is called slip casting.

Specific Gravity — a measure of mass relative to volume. The specific gravity of a metal is used to calculate the final weight of a casting.

Spiral Sawblade — a small blade used to cut wax and similar soft materials.

Spirit Lamp — see Alcohol Lamp.

Sprue — a wax rod, and later a passage through which molten metal is fed to a casting.

Spure Base — a pad, often of rubber, that makes a bottom for a flask during mold making. The cone or hemisphere on a sprue base makes the recesss that will be the pouring basin.

Stirring Rod — a tool used to stir and skim molten metal. They may be made of wood, carbon or quartz.

Taper — see Draft.

Tempered Sand — sand prepared to a correct moisture content for casting.

Tufa (Tuff) — a soft natural rock composed of compressed volcanic ash that is used as a mold.

Vents — see Risers.

Vulcanization — the process of curing natural rubber with pressure and heat to make it tough and elastic. In mold making this is done in a device called a vulcanizer.

Waste Mold — a type of mold that can be used only once because it must be destroyed to get at the casting.

White Metals — a family of low melting alloys with a silvery color. Principle ingredients are often tin, zinc, bismuth and antimony.

Common sense is your best protection. Even safe operations can hurt people who abuse them, so remember that accidents don't happen only to the "other guy." If you feel uncertain about a tool, a chemical or a process, stop and get assistance. If you feel ill or dizzy, leave the process and go to another room. If the illness persists, see a doctor and contact your state hospital system, Department of Occupational Safety.

Compound	Notes	Precautions
Acetone	Headache, drowsiness, skin irritation. One of the least toxic solvents.	Adequate ventilation.
Acetylene	Mild narcotic in small doses. Large doses cut off oxygen.	Use caution, check equipment for leaks.
Ammonia	Irritant to the eyes, caustic to lungs. Serious when in strong solution.	Use diluted soap and water.
Aqua Regia	Most caustic of all acids.	Wear protective clothing. Store in a well-marked loosely stoppered bottle.
Asbestos	Made up of fibers the body cannot dissolve. A carcinogen whose effects can take 20-30 years to develop.	Avoid it; use substitutes.
Benzene	Intoxication, coma, respiratory failure.	Avoid it; use an alternate solvent.
Cadmium	Affects brain, nervous system, lungs, kidneys.	Avoid if possible. If not. use only with strong ventilation.
Chlorinated Hydrocarbons	Dissolves the fatty layer of skin. Causes liver and kidney damage.	Avoid if possible. Ventilate, wear neoprene rubber gloves.
Copper Compounds	Oxides can irritate lungs, intestines, eyes and skin.	Ventilate when heating copper alloys, wear gloves when handling a lot, e.g. raising.
Cyanides	Mists inhaled or falling on skin are poisonous.	Ventilate well, wear protective clothing.

Fluorides	Can form hydrofluoric acid in the lungs.	Ventilate, avoid breathing the fumes.
Lead	Damages brain, central nervous system, red blood cells, marrow, liver, kidneys. Fumes are especially dangerous.	Avoid if possible. Ventilate well. Minimize handling and wash well after contact.
Ketones	Skin, eye and respiratory tract irritants. Can cause peripheral nerve damage.	Ventilate, wear rubber gloves and appropriate respirator.
Liver of Sulfur	When heated to decomposition it can create hydrogen sulfide, a source of brain damage and suffocation.	Do not allow mixture to come to a boil.
Mercury	Damages brain, nervous system and kidneys.	Avoid fumes and skin contact. Ventilate very well and wear protective clothing.
Polyester Resins	Skin irritants. Some release toxic fumes when mixed with their binders, some are explosive.	Wear gloves and ventilate. Store according to directions.
Silver Compounds	Absorbed into the skin as vapor or dust, can cause night blindness.	Wear gloves and a respirator.
Sulfuric Acid and Sparex (sodium bisulfate)	Irritates skin and respiratory tract. Damages clothing	Ventilate. Keep container covered. Neutralize with solution of baking soda and water.
Tellurium	Fumes generated in refining gold, silver, copper, and in welding. Irritates skin and gastro-intestinal system.	Ventilate. Early symptom is "garlic breath" and a metallic taste in the mouth. Be alert for this.
Toluene & Toluol	Causes hallucination, intoxication, lung, brain and red blood cell damage.	Avoid if possible. Ventilate well.
Turpentine	Skin irritant, possible brain and lung damage.	Ventilate and wear gloves.
Zinc Compounds	Dust and fumes attack the central nervous system, skin and lungs.	Ventilate and wear a respirator.

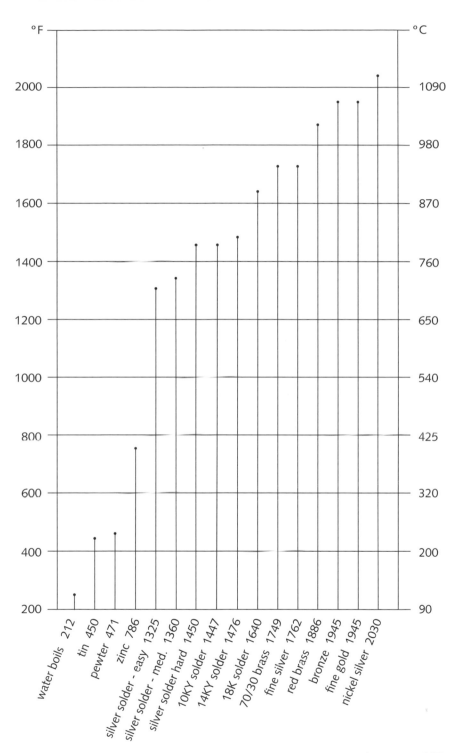

°F — water boils 212, tin 450, pewter 471, zinc 786, silver solder - easy 1325, silver solder - med. 1360, silver solder hard 1450, 10KY solder 1447, 14KY solder 1476, 18K solder 1640, 70/30 brass 1749, fine silver 1762, red brass 1886, bronze 1945, fine gold 1945, nickel silver 2030 — °C

	Metal or alloy	Au	Ag	Cu	Zn	Other	Melting Point		Sp.Grav.
Al	Aluminum					100 Al	660°C	1220°F	2.7
Sb	Antimony					100 Sb	631	1168	6.6
Bi	Bismuth					100 Bi	271	520	9.8
260	Cartridge brass			70	30		954	1749	8.5
226	Jewelers brass			88	12		1030	1886	8.7
220	Red brass			90	10		1044	1910	8.8
511	Bronze			96		4 Sn	1060	1945	8.8
Cd	Cadmium					100 Cd	321	610	8.7
Cr	Chromium					100 Cr	1857	3474	6.9
Cu	Copper			100			1083	1981	8.9
Au	Gold (fine)	100					1063	1945	19.3
920	22K yellow	92	4	4			977	1790	17.3
900	22K coinage	90	10				940	1724	17.2
750	18K yellow	75	15	10			882	1620	15.5
750	18K yellow	75	12½	12½			904	1660	15.7
750	18K green	75	25				966	1770	15.6
750	18K rose	75	5	20			932	1710	15.5
750	18K white	75				25 Pd	904	1660	15.7
580	14K yellow	58	25	17			802	1476	13.4
580	14K green	58	35	7			835	1535	13.6
580	14K rose	58	10	32			827	1520	13.4
580	14K white	58				42 Pd	927	1700	13.7
420	10K yellow	42	12	41	5		786	1447	11.6
420	10K yellow	42	7	48	3		876	1609	11.6
420	10K green	42	58				804	1480	11.7
420	10K rose	42	10	48			810	1490	11.6
420	10K white	42				58 Pd	927	1760	11.8
Fe	Iron					100 Fe	1535	2793	7.9
Pd	Lead					100 Pb	327	621	11.3
Mg	Magnesium					100 Mg	651	1204	1.7
	Monel Metal			33		60 Ni, 7 Fe	1360	2480	8.9
Ni	Nickel					100 Ni	1455	2651	8.8
752	Nickel silver			65	17	18 Ni	1110	2030	8.8
Pd	Palladium					100 Pd	1549	2820	12.2
	Old pewter					80 Pb,20 Sn	304	580	9.5
Pt	Platinum					100 Pt	1774	3225	21.4
Ag	Silver (fine)		100				961	1762	10.6
925	Sterling		92½	7½			893	1640	10.4
800	Coin silver		80	20			890	1634	10.3
	Mild steel					99 Fe, 1 C	1511	2750	7.9
	Stainless steel					91 Fe, 9 Cr	1371	2500	7.8
Sn	Tin					100 Sn	232	450	7.3
Ti	Titanium					100 Ti	1660	3020	4.5
Zn	Zinc				100		419	786	7.1

EQUIVALENT NUMBERS

B&S	mm	inches		drill size
		decimal	fraction	
0	8.5	.325	21/64	
1	7.34	.289	9/32	
2	6.52	.257	¼	
3	5.81	.229	7/32	1
4	5.18	.204	13/64	6
5	4.62	.182	3/16	15
6	4.11	.162	5/32	20
7	3.66	.144	9/64	27
8	3.25	.128	1/8	30
9	2.90	.114		
10	2.59	.102		38
11	2.31	.091	3/32	43
12	2.06	.081	5/64	46
13	1.83	.072		50
14	1.63	.064	1/16	51
15	1.45	.057		52
16	1.30	.051		54
17	1.14	.045	3/64	55
18	1.02	.040		56
19	0.914	.036		60
20	0.812	.032	1/32	65
21	0.711	.028		67
22	0.635	.025		70
23	0.558	.022		71
24	0.508	.020		74
25	0.457	.018		75
26	0.406	.016	1/64	77
27	0.355	.014		78
28	0.304	.012		79
29	0.279	.011		80
30	0.254	.010		

These factors allow you to calculate the weight of a known object in an alternate metal, as in

"How much would this sterling ring weigh in 18K gold?"

To change ⟶ Sterling ..... to	18K gold ... multiply by	1.48
	14K gold	1.248
	10K gold	1.104
	platinum	2.046
	fine silver	1.015

To change ⟶ Brass .... to	18K gold .... multiply by	1.885
	14K gold	1.589
	10K gold	1.406
	fine silver	1.273
	sterling	1.21

To change ⟶ 18KY Gold... to	18KW gold ... multiply by ...	1.064
	14K gold	0.842
	10K gold	0.745
	platinum	0.727
	sterling	0.675

To change ⟶ 14KY Gold ... to	18K gold .... multiply by	1.157
	14KW gold	1.035
	10K gold	0.884
	fine silver	0.791
	sterling	0.801

To change ⟶ Platinum .... to	18K gold .... multiply by	0.722
	14K gold	0.625
	10K gold	0.528
	fine silver	0.494
	sterling	0.483

INDEX

Air bubbles, 35,37-40

Alcohol lamp, 11

Alginate, 105

Alloys, 89,149

Ammonium chloride, 52

Baking soda, 129

Balancing, casting machine, 54

Bezel implants, 139

Binding wire, 74,137

Borax, 52

Bottom-pouring ladle, 90

Brazing, 131

Bronze molds, 100

Burnout, 41-47, 120

Burs for flexible shaft, 5, 132

Button, 56, 94

Carbon residue, 47

Centrifugal casting, 53-55,58

Chaplets, 120, 137

Charcoal, 70

Cope, 81-84

Core, 104,113,120,125,136

Cradle, 54

Crucible, 55, 67, 68, 127

Cup (pouring basin), 113-115

Cut-off needle, 12,17,26,73

Cuttlefish, 72-76

Debubblizer, 35

Dental tools, 6, 12

Double-metal, 133-134

Draft (taper), 78

Drag, 81-84

Extruder, 15

Electric wax pen, 13

Exhaust fan, 46

Fillet, 25, 26

Finishing, 9,17

Flame types, 49

Flashpoint, 2

Flask, 22, 24, 43

Flexible shaft burs, 5, 132

Flux, 52, 57, 126

Foundry sand, 77-80

Furnaces 42,50,51,122-124

Gate, 113-115

Gems, 140

Glossing off, 134

Glycerine, 80

Graphite crucibles, 50

Gravity casting, 94

Hard core (shell) investing, 40

Hydration, 34

Implants, 139

Ingate, 113,114

Inlay, 98

Investment, 31-40

Kilns, 45, 50, 51,121-124

Ladles, 67,90

Liquidus, 3

Locks, 103

Mandrels, 16

Mold frame, 81-84

Molding boards, 82, 85

Mother mold, 110

Cut out these tabs, fold each one over and paste it onto the first page of each chapter. Line up the top of the tab with the mark you'll find there.

1 Modelmaking	2 Sprue Systems
1 Modelmaking	2 Sprue Systems

3 Investing	4 Burnout &Melting
3 Investing	4 Burnout &Melting

5 Throwing	6 Direct Methods
5 Throwing	6 Direct Methods

7 Sand Casting	8 Pewter Casting
7 Sand Casting	8 Pewter Casting

9 Flexible Molds	10 Foundry
9 Flexible Molds	10 Foundry

11 Special Cases	12 Reference
11 Special Cases	12 Reference